AN ARCTIC WHALING DIARY
The Journal of Captain George Comer in Hudson Bay 1903–1905

When the American whaler *Era*, George Comer, Captain, sailed from New Bedford, Mass, for Hudson Bay in the spring of 1903, some American newspapers warned that there might be forcible intervention by the Canadian government, for the expedition coincided with sudden alarm about the precarious state of Canadian sovereignty in the Arctic and the dispatch of the government steamer *Neptune* to assert authority over a region which had been the preserve of American whalers for over forty years. In the end common sense prevailed, for, although the *Era's* activities were closely watched by the *Neptune*, which wintered by her in the harbour ice near Cape Fullerton in 1903–4, and by the *Neptune's* relief vessel the *Arctic* in the winter of 1904–5, and although relations between government personnel and whalemen were not always harmonious, the *Era* was able to follow her usual whaling procedures, seeking bowhead whales from May to September in both years.

George Comer, an experienced and skilful whaleman, was a disciplined recorder of daily events during all his whaling cruises and a vigilant and interested observer of the arctic environment and its native inhabitants. He compiled population figures, collected artifacts, photographed clothing and tattoo patterns of the various Inuit groups who participated in the whaling operations, and made the first wax cylinder recordings of their songs and tales. His journal of the 1903–5 expedition gives a valuable and fascinating insight into the arctic whaling industry, the lives of the native people associated with it, and the beginnings of Canadian intervention in the area. Professor Ross enhances this information with an introduction, epilogue, and notes, which describe Comer's career and whaling, and explain and enlarge on references in the diary.

W. GILLIES ROSS is a member of the Department of Geography, Bishop's University, Lennoxville, Quebec.

EDITED BY W. GILLIES ROSS

An Arctic Whaling Diary

THE JOURNAL OF
CAPTAIN GEORGE COMER
IN HUDSON BAY
1903–1905

UNIVERSITY OF TORONTO PRESS
Toronto   Buffalo   London

© University of Toronto Press 1984
Toronto Buffalo London
Printed in Canada
ISBN 0-8020-5618-0

Canadian Cataloguing in Publication Data

Comer, George.
An Arctic whaling diary
Bibliography: p.
Includes index.
ISBN 0-8020-5618-0
1. Comer, George, 2. Whalemen – Connecticut –
Biography. 3. Whaling – Hudson Bay – History –
20th century. 4. Hudson Bay – Description and travel.
I. Ross, W. Gillies (William Gillies), 1931-
II. Title.
SH383.2.C65   639′.28′0924   C83-098623-5

COVER: *Era* frozen into the ice at Fullerton Harbour, Hudson Bay   Photograph by George Comer   Mystic Seaport Museum

FRONTISPIECE: George Comer on rigging of the *Era*   Mystic Seaport Museum

This book has been published with the help of a grant from the Social Science Federation of Canada, using funds provided by the Social Sciences and Humanities Research Council of Canada.

Contents

PREFACE vii
ACKNOWLEDGMENTS x
MAPS xii
PHOTOGRAPHS xiv
INTRODUCTION 3

1 Voyage north
 (29 June – 23 September 1903) 40

2 Taking up winter quarters
 (24 September – 31 December 1903) 63

3 The first winter
 (1 January – 10 May 1904) 86

4 Spring and summer whaling
 (11 May – 25 September 1904) 116

5 Preparing for winter
 (26 September – 31 December 1904) 143

6 The second winter
 (1 January – 9 May 1905) 162

7 The second summer
 (10 May – 8 September 1905) 188

8 Voyage home
  (9 September – 15 October 1905) 210

EPILOGUE 218

APPENDIXES 225
A The schooner *Era* 225
B Crew list of the *Era* 1903 228
C Stores carried on the *Era* 1903 229
D Beaufort wind scale 235
E Minimum air temperatures at Fullerton Harbour 237
F Sea ice thickness at Fullerton Harbour 238
G Details of whales killed in 1905 240
H Whaling and trading returns 1905 241
I Population of Eskimo groups 243
J General observations by George Comer 245
K Government expeditions 1903–5 251

GLOSSARY 255
SELECTED BIBLIOGRAPHY 262
INDEX 265

Preface

'Born to ride out any storm, he sailed the seven seas, toughened by simple living, hard knocks and heavy work, from cabin boy to captain. George Comer, master of himself, became master of his ship with the muscles of a Hercules and a will of iron.'
Frederick Walcott at the dedication ceremony, East Haddam, Conn, 4 June 1938.[1]

It was a year or so after George Comer's death. Beneath a state fire tower on Mount Parnassus Road, a few miles east of the Connecticut River and the quaint town of East Haddam, over 200 people had gathered on Saturday 4 June 1938 for the unveiling of a granite tablet in memory of the whaling captain. The site was close to 'Restwood,' then as now the home of the Comer family. Members of the Connecticut Forest and Park Association were hosts at the ceremony and music was provided by the Fife and Drum Corps of nearby Moodus. Among those who spoke were: William L. Cross, governor of Connecticut; Frederick C. Walcott, Connecticut state welfare commissioner and former United States senator; Robert Cushman Murphy, ornithologist at the American Museum of Natural History in New York; Mrs Edward Stafford, a daughter of the polar explorer Robert E. Peary; and Charles E. White, president of the Savings Bank of New London. Comer's grandson and namesake, then aged twenty-one, unveiled the tablet which read 'To the memory of Captain George Comer, able seaman, arctic mariner, navigator of the seven seas, 1858–1937.'

---

1 The typescripts or newspaper accounts of remarks made at the dedication ceremony are in the George Comer Papers, East Haddam, Conn, 3–8.

The dedication of a memorial, the participation of state and local institutions, and the presence of a number of well-known people from government, science, and commerce suggest that George Comer was no ordinary man. Indeed he was not. The diverse activities and achievements of his seventy-nine years had touched a wide spectrum of acquaintances and admirers, many of whom were privileged to regard him as friend. Whaling had been his business, but he had never limited himself to the task of navigating small ships through icy waters in search of leviathan – a task which by itself would appear to offer more than enough challenge for anyone. George Comer 'could not.endure idleness' as LeRoy Harwood, treasurer of the Mariners Savings Bank of New London said at the ceremony, and to the demanding routine of polar ship handling, sealing, and whaling, he added the study of polar animals and birds, the observation of matters of geographical interest, and the investigation of Eskimo culture. His insatiable curiosity, extensive experience in the polar regions, and ability as a collector brought him into contact with some of the foremost explorers and scientists of the day. He corresponded with Roald Amundsen, the first man to pilot a ship through the North-west Passage, Vilhjalmur Stefansson, advocate of the 'Friendly Arctic' concept, Knud Rasmussen, the Danish explorer who sledded from Hudson Bay to Bering Strait during the Fifth Thule Expedition of 1921–4, Franz Boas, author of the first major ethnographic work on the Canadian Eskimo, and the Reverend E.J. Peck, who introduced syllabic writing to the Eskimos of Baffin Island. At the time of his death on 27 April 1937, and a year later at the dedication of the memorial, those who praised George Comer spoke not only of his success in the whaling trade, but also of his contributions to geography, zoology, archaeology, and anthropology, and of his participation in community and state affairs. Above all they spoke of his character as a man.

A portrait of George Comer in his later years shows a rather rotund face with a moustache which gives a whimsical, slightly walrus-like appearance. The eyes reveal sincerity and warmth, and even hint at humour, but at the same time suggest a certain firmness. A man, it would appear, who might be friendly rather than aloof or antagonistic, open rather than secretive, tolerant rather than narrow-minded, and yet a man whom one would be anxious not to disappoint or deceive, a man with a sense of right and wrong who would set high standards both for himself and for those under his authority. He was a big man, a shade under six feet and weighing about 200 pounds, conspicuous by his size and bearing. People were naturally drawn to him and many whalemen were ready to accompany him year after year to some of the harshest whaling grounds in the world. Brass Lopes, a boatsteerer on the 1903–5 voyage, was one. Another, Frank Borden, wept as the *Era* departed for Hudson Bay because poor health prevented him from going.

Comer's friendliness and honesty made him good friends among Aivilingmiut and East Haddamites alike. In Hudson Bay, where the ship was his home during nine long months of winter, the Eskimos were always made welcome. As he records in his journal (10 March 1905), 'all natives have the freedom of the cabin to come and go when they like.' At 'Restwood' too visitors were always welcome. His 'heart and home were symbols of hospitality,' as Robert Cushman Murphy said at the dedication. Visitors might be shown some carved walrus tusk cribbage boards or other mementos brought back from the Arctic on one or another voyage, and might be treated to a few stories of adventure in the far north. Jovial with a good sense of humour, Comer was an entertaining raconteur and many organizations invited him to speak about his polar voyages. The Explorers' Club of New York, the Jibboom Club of New London, the Middletown Chamber of Commerce, the Children's Museum of East Haddam, and a number of other institutions both large and small, national and local, benefited from his informative lectures, usually illustrated with lantern slides made by the captain himself.

Always conscious of 'simple human relationships,' as Murphy noted, George Comer was not only hospitable but also generous and considerate. When he sent apples to the scientists at the American Museum of Natural History he took care that some should reach the secretaries as well. His kindness was reciprocated. Residents of East Haddam and New London often sent packages with him on whaling voyages to open when he pleased. The periodic unwrapping of some surprise gift from friends at home punctuated the loneliness and discomfort of dark arctic winters, and brought him much pleasure.

At the memorial dedication his long-time friend LeRoy Harwood summed him up: 'Big in body, strong in mind, a strict disciplinarian, but a sympathetic friend of all mankind, he represented in his life all that is best in human character.'

W.G.R.

Acknowledgments

The editing of George Comer's personal record of an arctic whaling voyage has been a pleasant side trip from research into the development and impact of commercial whaling in Canada's Arctic. My particular interest in Comer, and in the voyage of 1903–5, arose during a study of the influence of whaling upon the Eskimo inhabitants of the Hudson Bay region. That work was assisted by the Canada Council, the National Museum of Man, the Department of Indian Affairs and Northern Development, the Advisory Committee on Geographic Research, the Canadian Research Centre for Anthropology (all in Ottawa), and Bishop's University (Lennoxville, Quebec). I am pleased to acknowledge the generous support of these institutions, support that led indirectly to this book.

In preparing the introduction, notes, and appendixes I utilized several libraries, archives, and museums. I thank the staffs of all, particularly those of the Public Archives of Canada and the Scott Polar Research Institute, Cambridge, England. Many individuals helped with specific problems or in special ways. Among them are: Terence Armstrong (Scott Polar Research Institute, Cambridge); Frederica de Laguna (Bryn Mawr College, Bryn Mawr, Penn); Alfred Copland (Ottawa); Bernadette Driscoll (Winnipeg Art Gallery); Linda Eisenhart (Smithsonian Institution, Washington); Basil Greenhart (National Maritime Museum, Greenwich); Horst Hartmann (Museum für Völkerkunde, Berlin); Richard Kugler (Old Dartmouth Historical Society, New Bedford, Mass); Trevor Lloyd (Ottawa); Claude Minotto (Public Archives of Canada, Ottawa); Benoit Robitaille (Université Laval, Quebec); Stuart Sherman (John Hay Library, Brown University, Providence, RI); A.N. Stimson (National Maritime Museum, Greenwich); and Garth Taylor (National Museum of Man, Ottawa). The anonymous readers for the University of Toronto Press made valuable suggestions concerning the manuscript. The photographs have been reproduced with the help of a grant from the Publications Committee of Bishop's University.

The aerial photographs, © 1956 Her Majesty the Queen in Right of Canada, are reproduced from the collection of the National Air Photo Library with permission of Energy, Mines and Resources Canada.

I acknowledge the co-operation and assistance of the archival staff of the G.W. Blunt White Memorial Library at Mystic Seaport, Mystic, Conn, where Comer's 1903–5 journal is, and the careful work of the Seaport's photographic section which provided several of the illustrations for this book from copy negatives made from Captain Comer's own glass plate negatives.

I am grateful also to George Comer, grandson of the whaling captain, with whom I have enjoyed several 'gams' in East Haddam.

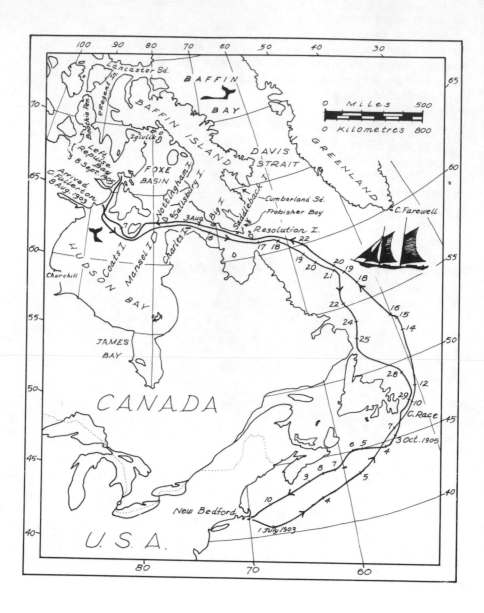

Era's track to and from Hudson Bay

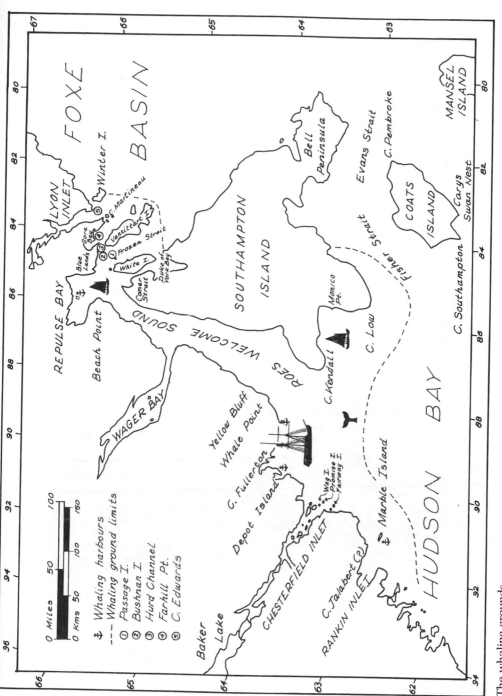

The whaling grounds

Captain George Comer
Mystic Seaport Museum

The crew of the *Era* in Hudson Bay   The Whaling Museum, New Bedford,
Mass

The schooner *Era* under sail   Mystic Seaport Museum

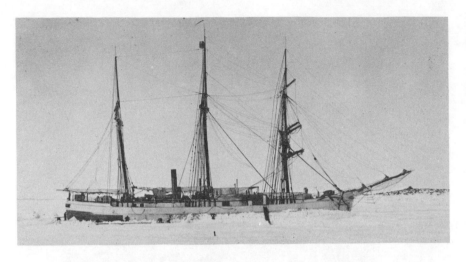

The government steamer *Arctic*   Stefansson Collection, Dartmouth College

The crew of the *Neptune* on the banking around the hull   Photograph by
J.D. Moodie   PAC PA-53567

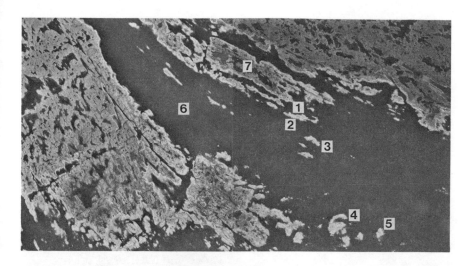

Winter quarters in Fullerton Harbour between Bernheimer Bay (6) and Cape
Fullerton on the east. Located at the foot of Major Island (7), the harbour was
protected by Store Island (2), and the approach was marked by beacons on
Beacon Island (3), Barrel Island (4), and Outer Barrel Island (5)
National Air Photo Library, photos A-15409-4 and A-15410-30

Summer anchorage (1) in the Harbour Islands, Repulse Bay   National Air
Photo Library, photos A-15350-182 and A-15352-31

The Scottish steam whaler *Active*   Dundee Corporation, Art Galleries and Museums, Albert Institute, Dundee

Hauling pond ice   The Whaling Museum, New Bedford, Mass

Eskimos in snow house   Photograph by George Comer
Stefansson Collection, Dartmouth College

Eskimo costumes drawn by Meliki, an Aivilik Eskimo   Courtesy of the
American Museum of Natural History

Ready for a facial cast (left) and taking the cast (right) Photographs by George Comer Courtesy of the American Museum of Natural History

 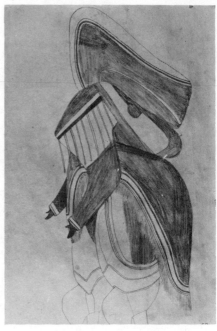

Tom Luce, unofficial mascot of the *Era*, named after the vessel's owner Photograph by George Comer Mystic Seaport Museum

Eskimo drawing by Meliki, an Aivilik Eskimo Courtesy of the American Museum of Natural History

Communication between the shaman and the soul of a sick person, established through a third person whose head could not be lifted if the soul wished to reply affirmatively to a question from the shaman but could be lifted easily if the response were negative   Photograph by George Comer   Mystic Seaport Museum

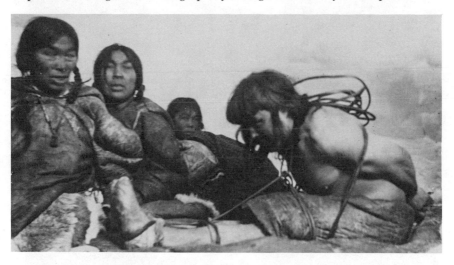

After being bound in thongs, the shaman could manage to free himself and thereby temporarily release souls from their bodies
Photograph by George Comer   Mystic Seaport Museum

Photograph by Grace Moodie   British Museum

Traditional women's costumes elaborately decorated with manufactured cloth
and glass beads obtained through trade with whaling vessels
Photograph probably by Grace Moodie   RCMP Archives

Shoofly sewing in the 'house' on *Era*'s
deck   Photograph by George
Comer   Mystic Seaport Museum

Shoofly in gala dress
Photograph by A.P. Low
PAC PA-53548

OPPOSITE: Hattie and Jennie dressed for a dance in clothing imported for them
by the whalemen   Photograph probably by Grace Moodie   RCMP Archives

Preparing for spring whaling   Photograph by A.P. Low   PAC PA-38270

Two whaleboat crews stand on the ice in front of their covered boats
Photograph by A.P. Low   PAC C-46972

A whaleboat returning from a cruise to Southampton Island pulled up onto the ice for the night and festooned with a polar bear skin, dead foxes and geese, caribou antlers and skin, guns, drying clothes, and baggage
Photograph by A.P. Low   PAC PA-53582

Crew inside covered boat   Photograph by George Comer
Mystic Seaport Museum

Crew, whaleboat, and *Era* at Fullerton   Photograph by A.P. Low
PAC PA-53568

Beached whale   Photograph by George Comer   Mystic Seaport Museum

Monday   July 31   1905
Lyons Inlet
A very heavy gale from south
with rain   at the tide gone
low this afternoon   we got it
Bone was 6 feet 10 in
the whole natives are working
with us   saving the meat and Blubber
at night weather clear a little

Tuesday   August 1
Some fog   and rain   we
were scraping Bone till 3 h m
when a whale was raised
pushed off   and saw two
Mr Reynolds got fast to the
smaller one   got it   ashore
at 6 h m   low tide
the whale was saved by a
native boy living here
later he got his box
of Tobacco   this is a
prize given to the Person
who first sees a whale and it
is caught
length of Bone 6 feet 10 inches

A page from Comer's journal   Mystic Seaport Museum

Spreading ashes on the ice in lines radiating from the vessel to accelerate the melting of harbour ice in spring   Photograph by George Comer
Mystic Seaport Museum

Whaleboats towing *Era*   Photograph by George Comer   Mystic Seaport Museum

AN ARCTIC WHALING DIARY

Introduction

This journal by George Comer, master of the American whaling schooner *Era* in 1903–5 and probably the best known of all Hudson Bay whalemen, provides a vivid picture of the arctic whale fishery in its final phase. For almost half a century whaling ships – mainly American – had been visiting Hudson Bay in search of bowhead whales, and for more than two centuries the vessels of several European nations had exploited other whaling grounds between Greenland and Baffin Island. Operating beyond British and Canadian authority the whaling captains had plied their ships freely through the hazardous but profitable waters, killing whales and other animals, and employing and trading with the native people. The first substantial exploitation of resources by outside interests and the first major thrust of an alien culture among the Eskimos, whaling was a powerful ecological force in the sensitive northern regions and the precursor of many economic and social developments in the Canadian Arctic. It brought about fundamental changes in human geography and played a role in the extension of Canadian sovereignty and administration into the region.

By 1915, when economic and biological factors brought the arctic whaling industry to a close, there had been well over 6,000 voyages into the waters between Greenland and the central Canadian Arctic.[1] Most were to the Davis Strait whaling grounds, the vast area encompassing Davis Strait, Frobisher Bay, Cumberland Sound, Baffin Bay, Lancaster Sound,

---

1 W. Gillies Ross, 'The Annual Catch of Greenland (Bowhead) Whales in Waters North of Canada 1719–1915: A Preliminary Compilation,' 106.

and Prince Regent Inlet, while Hudson Bay, the immense inland sea that extends into the heartland of the Canadian north, attracted fewer than 150 voyages. None the less, Hudson Bay whaling was important because ships normally overwintered during their voyages and an unusually intense and sustained contact developed between whalemen and Eskimos. In addition, when the Canadian government decided at last to supervise and regulate the activities of foreign whalers in territorial waters, it concentrated its initial efforts in Hudson Bay, precisely at the time of Comer's voyage on the *Era* in 1903.

Arctic waters support several species of sea mammals of which the largest is *Balaena mysticetus*, popularly known as the Greenland whale, polar whale, right whale, common whale, or bowhead. Now scarce owing to the depredations of the whalers, this magnificent cetacean was formerly abundant in the eastern Arctic. One major population, in the order of 11,000 in 1825, migrated northward in summer from its wintering grounds between Labrador and Greenland through Davis Strait into Baffin Bay and the Lancaster Sound region.[2] A second, possibly discrete population of perhaps 700 migrated through Hudson Strait into Hudson Bay.[3] The bowhead of the whaling period was rotund and up to fifty-five feet long. Its cavernous mouth had several hundred narrow plates of springy baleen hanging from its upper jaw and enabling it to strain small crustaceans out of sea water for food. These slabs of whalebone, some of which could exceed twelve feet in length in a large whale, were used commercially for making riding crops, buggy whip handles, umbrella ribs, corset stays and other items requiring resilience. The hair-like fibres along the trailing edge of the slabs were used in brushes and as stuffing for upholstery. Until the advent of spring steel and plastic, whalebone was in great demand. The bowhead also had a generous layer of blubber to restrict heat loss in the cold northern seas, and this yielded immense quantities of oil to lubricate machines and illuminate towns and cities before the advent of petroleum and coal gas.

Because the kill of one large bowhead whale could provide fifty barrels of oil and a ton of whalebone, industry began systematically to exploit the species in arctic waters in the early seventeenth century around Spitsbergen, and later extended westward in succession to the East Greenland Sea, the Davis Strait region, Hudson Bay, the Sea of Okhotsk, Bering Sea, and finally the Beaufort Sea.

2 Edward Mitchell and Randall R. Reeves, 'Catch History and Cumulative Catch Estimates of Initial Population Size of Cetaceans in the Eastern Canadian Arctic.' Report to the International Whaling Commission, 1979 (sc/32/016), 9.
3 Ibid. 10.

Davis Strait Whaling

European whaling first extended into the waters between Greenland and Canada in the seventeenth century, when the Danes established shore stations along the west Greenland coast and Dutch and German vessels began cruising northwest of Cape Farewell to trade with Greenland Eskimos and catch whales. American ships reached Davis Strait by 1729 and British whaleships around 1750, but it was the Dutch who dominated the Davis Strait whale fishery during the eighteenth century. After 1800, however, ships from England and Scotland made up most of the fleet, along with a handful of Dutch ships in the 1820s and a number of American whalers after 1845.

At first the ships whaled mainly along the west coast of Greenland where bowheads were commonly found during the months of April, May, and June. After 1820 they exploited the coast of Baffin Island as well – the 'West Side' – and extended the duration of their voyages to September. In a typical voyage of the 1830s a vessel would ascend the Greenland coast in May and June as the whales moved north towards their summer range, cross through Melville Bay to the mouth of Lancaster Sound by mid-July, cruise near Pond Inlet for a few weeks, then work southward along the Baffin Island coast with the whales' return migration in August, and head homeward in September. In some seasons, of course, this general pattern would be altered in response to variations in the distribution of sea ice. The whaling was more or less pelagic, carried out by whaleboat crews while the vessels were under way, moored to the edge of landfast ice, or anchored in bays. But as whalemen became more familiar with the seasonal migration of the whales and the geography of the region, they established contacts with the land and made good use of the considerable skills of the native population, particularly on the West Side.

The focal point of these developments was Cumberland Sound, a broad gulf penetrating 150 miles into the southeastern part of Baffin Island. The region had several advantages. It abounded in whales, particularly in spring and late autumn; it was less affected by pack ice and icebergs than the water beyond its mouth; the whaling season continued late into the fall; and it supported a large Eskimo population – perhaps as many as a thousand.[4] After being introduced to the Sound by an Eskimo pilot in 1839 British whalemen were quick to recognize its possibilities, and within a few years Cumberland Sound had become one of the principal centres of whaling activity in the arctic regions. Some whaling masters – the Amer-

---

4 William Penny to Captain Beaufort, 15 October 1840. Scott Polar Research Institute, Cambridge (MS 116/63/46).

icans in particular – abandoned the traditional counter-clockwise circuit of Davis Strait and Baffin Bay in favour of sailing directly to Cumberland Sound.

The native population there was quickly attracted into a close economic and social relationship with the whaling crews. They manned whaleboats, pursued whales, participated in flensing, transported blubber by dog sled, hunted to provide fresh caribou meat, acted as guides on sled trips, made up and repaired skin clothing, and carried out many other tasks. These beneficial ties between whalemen and Eskimos, which steadily grew more systematized and more intense, were a major characteristic of whaling in this region.

In addition to the utilization of native labour, whaling in Cumberland Sound was characterized by the practice of wintering. Until 1851 voyages to the Davis Strait whale fishery lasted only one season unless ships were trapped in the ice before they could depart homeward, but in that year an American whaling master left fourteen of his crew to winter among the Eskimos of Cumberland Sound. The experiment proved highly successful, the whalemen being able to continue their operations later in the fall and resume them earlier in the spring. Two years later three crews, two British and one American, overwintered on board their vessels frozen into harbour ice. Soon this became a relatively common technique for extending the duration of whaling on Davis Strait voyages. Because wintering success depended to a considerable extent on Eskimo assistance in making cold weather clothing and procuring fresh meat, it required harmonious and regular relationships with the native people, and led inevitably to profound changes in their way of life.

A third characteristic of Cumberland Sound whaling, made possible by overwintering and employing native labour, was the practice of floe whaling, that is, whaling from the floe edge, in spring. When whales migrated into Cumberland Sound in May after spending the winter in Davis Strait, their progress was restricted by the landfast ice which stretched solidly from shore to shore over most of the Sound. They therefore tended to congregate at the floe edge (the seaward margin of this ice cover), advancing gradually up into the gulf as the ice melted and disintegrated, their range expanding as fast as the floe edge retreated. From winter quarters around the periphery of Cumberland Sound, the whalemen and their native helpers moved whaleboats and gear to the floe edge by dog sled in May, camped there, and launched their boats from the ice edge whenever whales appeared. After a kill they flensed the carcass in the water, and sled drivers transported the bone and blubber back to the ships to store or boil down. Floe whaling was a privilege of wintering vessels; incoming ships could not penetrate the Davis Strait pack ice so early in the season to enter Cumberland Sound.

Hudson Bay Whaling
Whaling might never have begun in Hudson Bay without the benefit of experience in Cumberland Sound where the techniques of wintering, floe whaling and utilizing native labour had developed in the 1840s. In the eighteenth century there had been a brief attempt by the Hudson's Bay Company to secure bowhead whales using trading sloops operating from its Churchill trading post, but the experiment had proved quite unsuccessful.[5] The whaling period in Hudson Bay really began in 1860, and arose directly out of American activities in Cumberland Sound. A voyage by two American whalers, the *Syren Queen* and the *Northern Light*, inaugurated Hudson Bay whaling. Captain Christopher Chapel and his brother Edward decided to investigate reports of many whales in northwest Hudson Bay. Their initiative paid off handsomely. When the vessels departed homeward in the fall of 1861, having wintered near Depot Island, the *Northern Light* carried 21,000 pounds of whalebone, more than any American ship had ever obtained in the Davis Strait fishery, and the *Syren Queen* contained 15,700 pounds.[6] News of this phenomenal catch attracted other whalemen to the region. In the decade from 1860 to 1870 there were fifty-nine voyages into Hudson Bay, all but two from American ports.[7]

The Hudson Bay whaling grounds were completely separate from those of the Davis Strait region and involved what appears to have been a different stock of whales. In summer whales were to be found in the northwest section of the Bay, ranging about 400 miles from Marble Island in the south through Roes Welcome Sound to Lyon Inlet in the north. From the North Atlantic through Hudson Strait and across the northern part of the Bay to this whaling ground was more than 900 miles, a journey almost half as long as that from New England ports to the mouth of Hudson Strait, and one in which incoming ships had to contend with heavy concentrations of pack ice and bergs, as well as adverse winds and, in places, contrary currents. A sailing vessel could count on reaching Resolution

5 W. Gillies Ross, 'Whaling in Hudson Bay. ı. The Marble Island Whale Fishery 1765–1772,' *Beaver*, outfit 303, no 4 (1973), 4–11.
6 Alexander Starbuck, *History of the American Whale Fishery from its Earliest Inception to the Year 1876* (1878; reprint, New York: Argosy-Antiquarian 1964), 2: 577.
7 W. Gillies Ross, *Whaling and Eskimos: Hudson Bay 1860–1915*, 37.

Island at the entrance to the Strait in late July, approximately a month out from New Bedford or New London, but another two weeks to a month were needed for the passage to Roes Welcome Sound, the heart of the whaling ground. By the time whalers arrived in mid-August, there remained only a month of cruising before autumn gales, low temperatures, fog, snow, and young ice would announce the approach of winter and the end of ship navigation. As a result, one-season voyages were scarcely worth the effort, and few captains bothered to make them. From the start of whaling in 1860 the usual practice was to enter the Bay as early as possible, cruise as long as feasible, spend the winter frozen into a protected anchorage where fresh water and game could be obtained, and resume whaling in the following spring, launching whaleboats from the floe edge in May in the Cumberland Sound tradition, long before the whaleship could be liberated from the harbour ice. In this way the short one-month whaling season of the first summer could be followed by a four-month season in the second.

Of the several locations suitable for the wintering of whaling vessels, Marble Island, at the southern extremity of the whaling ground, was one frequently used, especially in the early decades. It was easily accessible to ships, possessed a good anchorage and a landlocked inner harbour, and was close to the floe edge, well situated for spring whaling. On the other hand, Marble Island was often cut off from the mainland by unstable ice conditions which severely restricted Eskimo contact and hunting; a high incidence of scurvy was usually the result when meat could not be obtained. Other anchorages farther north, at Depot Island, Cape Fullerton, Repulse Bay, and Lyon Inlet, attracted wintering vessels from time to time, and provided occasional protection in summer to ships cruising for whales.

When a ship first arrived on the whaling ground, usually in August, her master would select a winter harbour, proceed there, and unload stores to be used later. A few crews of native whalemen might be taken on board to help with the whaling, or commissioned to whale independently on behalf of the vessel, and arrangements would probably be made for a number of Eskimo hunters and their families to work for the ship during the winter. Initial organization completed, the master would take the vessel out to cruise for whales in Roes Welcome Sound during the remainder of the navigation season. By late September, when cold weather made sailing, boat handling, and whaling unbearably difficult and hazardous, it was time to place the ship in winter quarters.

During summer whaling the normal procedure was for a ship to cruise leisurely up and down Roes Welcome Sound. Lookouts posted high above the deck, protected from cold winds by an enveloping crow's-nest, scanned the waters for the tell-tale blow of a whale at the surface. A cry from aloft would bring the boat crews racing on deck from their bunks, hurriedly dressing as they ran, and within minutes several whaleboats would be lowered in pursuit of the quarry. The boats were double-ended, from twenty-five to thirty feet long, propelled by oars or by sail, and generally manned by six men. As the sailors pulled on the oars, the boatheader at the stern, guided by signals from the ship, used his long steering oar to keep the boat on course towards the whale. When close, the bowman would ship his oar, stand facing forward, and take up his harpoon, a wooden-shafted, iron-headed weapon with a toggling point designed to turn sideways after penetration to prevent withdrawal. At the right moment he would drive the harpoon into the body of the whale. With this thrust the harpooner united boat, men, and whale in an exhausting and dangerous dance which sometimes endured for several hours. Its object was to weaken and tire the whale with the drag of the boat, so that the death blow could be delivered more safely. A boat crew had to play its immense prey as a sport fisherman plays a trout or tuna. From the harpooned whale the line extended to the whaleboat where it ran over the bow chocks, down the centre of the boat, around an upright loggerhead at the stern, and into line tubs on the floorboards. By hauling in line when slack and easing out when taut the men allowed the boat to be dragged behind the stricken whale without foundering or being pulled underwater. When the whale, weakened from loss of blood and fatigued from dragging the boat, slowed its frantic flight, the men could haul up to the animal and the boatheader would come forward to make killing thrusts into the body with a long, sharp lance.

The time taken to harpoon and kill a whale depended on its size, strength, and endurance, the conditions of weather and ice, the skill of the whalemen, the number of whaleboats involved, and the weapons used. Although hand-thrown harpoons and hand-held lances remained standard weapons until the end of arctic whaling, several alternative weapons were available in the nineteenth century and were employed in Hudson Bay. The Scots often used harpoon guns mounted on their whaleboats rather than hand harpoons. The Americans made effective use of the shoulder gun which fired a small explosive projectile into a harpooned whale, making the kill safer and quicker. Another weapon used widely by American whalers was the darting gun which performed the functions of both harpoon and lance. It consisted of a traditional harpoon with the equiv-

alent of a shoulder gun attached to its wooden shaft. Triggered auto-
matically when the harpoon entered the whale, the gun sent a bomb to
explode deep in the body, so that in one operation the animal could be
killed and attached to the whale line for retrieval. Even with weapons
such as these, however, several hours were often needed to subdue large
whales.

If a dead whale could not immediately be collected by the whaleship
the men faced the arduous work of towing the huge carcass with their
boats, a task which could become almost unendurable when weather and
seas conspired to obstruct their progress. Once at the vessel, the whale
was secured tail forward along the starboard side, and the men commenced the flensing, stripping the oil-filled blubber away in long rolls as
the body rotated in the water, hoisting the strips on deck to be cut into
manageable pieces, slicing them fine, and feeding small portions into the
tryworks for rendering.

Certain paraphernalia were required for the business of processing
whales: a cutting stage mounted outboard on which the men could stand
to dissect the floating carcass; various tools for slicing and lifting; heavy
tackle and blocks at the mast cross trees to take the weight of the blubber
as it was removed; the tryworks, which were twin metal cauldrons set
above fires fuelled by blubber scraps, entirely encased in brick and surrounded by a water pan to reduce the terrible danger of fire; and, lastly,
scores of wooden casks carried north as shakes and hoops and made up
when needed on the whaling grounds for the storage of oil or blubber.
British methods were slightly different. Flensing was done from a floating
boat tied alongside the whaler rather than from a cutting stage, and
blubber was usually transported to home port instead of being processed
on the whaling grounds.

When summer whaling drew to a close a master intending to remain
in Hudson Bay would take his ship into the harbour selected for wintering
and remain at anchor in a safe position until the vessel was securely held
in a continuous sheet of stable ice. The anchor could then be raised on
board, and the convenient solid surface of the harbour used to transport
spars, whaleboats, and extra stores ashore, making more living space on
the ship. Ice had to be cut out of ponds and brought to the vessel to melt
down later for drinking water. To create additional space for work or
social functions, part of the upper deck was normally closed in with lumber
brought north for the purpose. Snow was banked up around the hull for
insulation. These tasks were carried out before Christmas. After that the
rest of the winter could be passed in relative idleness. Wise captains kept

the men busy, however, and many wintering crews participated in classes, concerts, theatricals, dances, sporting events, sled travel, and hunting. The men frequently added to the ship's larder the flesh of birds, fish, caribou, musk-oxen, seals, and walrus, but the real burden of procuring food fell upon the native hunters employed by the ship, who went inland to hunt caribou during the autumn migration and at times during the winter months.

As the rising temperatures and lengthening days in April made outside activities more tolerable, preparations were begun for spring whaling and a sense of purpose would pervade the crew. At this time whaleships would still be held solidly by harbour ice six feet thick; they could seldom be extricated before mid-July unless the men undertook vigorous efforts with huge ice saws. As soon as migrating whales made their appearance among the drifting ice floes beyond the coastal fringe of fast ice, the whalemen would temporarily abandon their immobilized vessels, leaving only a few 'shipkeepers' on board, and would transport their whaleboats and gear by native dog sleds to tent camps on headlands near the floe edge or on the fast ice itself. From these bases they carried out their whaling, launching their boats from the ice, hauling them out when the pack ice closed in, and flensing dead whales at the floe edge. The weapons and flensing tools used were no different from those employed when boats were lowered from cruising vessels in summer whaling. The darting gun, however, was especially useful amid pack ice in spring; a harpooned whale had to be killed quickly before it could run beneath the ice and force the attached boat to cut loose (and lose the whale) or be dashed to pieces.

In the last decades of Hudson Bay whaling the spring routine was modified. It was no longer worthwhile for boat crews to wait passively at the floe edge for whales to appear nearby because the whales were by then too scarce. George Comer carried the offensive to the whales, seeking them out as in summer, but in whaleboats rather than whaleships. This was a bold innovation which required capable leadership and skilful boat handling. The crews would set off on two-week journeys, carrying whatever camping gear and food they needed, sailing their small open boats hundreds of miles in uncertain conditions of weather and pack ice, largely cut off from protective harbours by the barrier of landfast ice. They took refuge wherever they could, often hauling out on ice floes or shore ice to camp for the night. Their boats were as much their homes as vehicles of transportation and instruments of whale-killing; at night canvas covers could be erected over them to create cosy but cramped boat-tents in which the men ate and slept. When a whale was killed on such an expedition there was no question of saving bulky blubber, which by this time had become relatively worthless. The whalebone was cut out and saved but the rest of the carcass was left to rot.

Comer, in substituting long whaleboat cruises for the customary floe whaling operation, extended the geographical limits of spring whaling far beyond the vicinity of the winter harbours. He incorporated the concept of long-range boat expeditions into the summer routine as well, using the whaleship more as an itinerant base then as an active whaling vessel. In 1905, for example, Comer commanded long boat voyages in May and June while the *Era* remained frozen into harbour at Cape Fullerton; in July he took the ship out of harbour and sailed her to an anchorage in Repulse Bay, from which the boats again cruised widely, reaching as far east as Lyon Inlet at the northern extremity of the whaling ground.

The Decline of Whaling

When the Hudson Bay grounds were first exploited in 1860 by the two American vessels whaling was a major industry in the United States. More than 500 American whalers cruised the world's oceans in that year, most seeking the sperm whale of warmer waters.[8] But the apogee of whaling had passed. A number of factors contributed to the decline that would bring the industry to collapse within half a century. Men left the whaleships to seek gold in California or to work in the cotton mills in New England cities. Civil War losses reduced the whaling fleet by half. Arctic ice took a heavy toll of ships, especially in the waters off Alaska. The growing scarcity of whales necessitated longer and more costly cruises, and capital migrated into safer industries with faster, more certain, returns. Coal gas became an alternative illuminant and petroleum products steadily replaced whale oil for both lighting and lubrication. The American whaling fleet fell from 736 vessels in 1846 to a mere forty in 1901.[9]

The decline would have been even more precipitous had it not been for the strong demand for baleen, which quadrupled in price between 1866 and 1904 (while the price of right whale oil dropped by more than three-quarters). A large adult bowhead could yield more than 2,000 pounds of bone so that as the price per pound exceeded $1.00 in 1863, $2.00 in 1876, $3.00 in 1884, $4.00 in 1890, and $5.00 in 1891, there was a strong incentive to outfit vessels for the arctic fishery.[10] The whalebone boom did not arrest

8 Walter S. Tower, *A History of the American Whale Fishery*, Series in Political Economy and Public Law no 20 (Philadelphia: University of Pennsylvania Press 1907), 121.
9 Ibid. 121.
10 Ibid. 128.

the general decline within the whaling industry, but it did extend for a few more decades the marginal operations of the few American, Scottish, and English firms still left in the business.

Arctic whaling around 1900 was a pale shadow of what it had been before mid-century. In 1830 more than ninety vessels had sailed to the Davis Strait whaling grounds, but in 1900 a mere half-dozen set out.[11] In several seasons during the peak period from 1820 to 1835 the British fleet had secured more than a thousand whales – indeed more than 1,600 in the year 1833 – but in 1900 the total catch amounted to fewer than twenty whales.[12] The story was similar in Hudson Bay. After fifty-nine voyages to the region during the first decade 1860–70 there were only twenty or so in each subsequent decade, an average of two a year.[13]

It is essential to appreciate that George Comer's Hudson Bay voyage of 1903–5 on the *Era* was carried out in the twilight of the arctic whaling period. But if his journal is not replete with thrilling accounts of encounters with bowhead whales, is it any less interesting as a narrative or less important as a historical document? In revealing to us that the mere sighting of a whale had become almost a rare event by 1903, Comer's journal presents a rather different portrait of arctic whaling. While perhaps an unexpected picture, it is none the less an accurate one, and it reflects the desperate adaptations made by a marginal industry to sustain itself as the whales became fewer and fewer.

Whales were still the primary objective of the voyage, and in its season whaling took precedence over all other activities. But, while owners still hoped to obtain large quantities of whale oil and bone, they also encouraged their captains to return with other animal products secured through systematic trade with the native people. The beginning of the fur and ivory trade can be traced to the casual barter for native handicrafts often carried out by the sailors on early voyages. This souvenir trade had developed into an organized, economically important activity in which the returns accrued to the owners of the vessels rather than to the individual whalemen. As whaling had become less profitable, enterprising whaling masters had encouraged the Eskimos to obtain the skins and furs of arctic fox, wolf, wolverine, caribou, musk-ox, polar bear, seal, and walrus, and the ivory tusks of walrus and narwhal. In Hudson Bay whaleships had begun trading commercially for furs as early as 1870, forty years before

11 Ross, 'Annual Catch,' Table 3.
12 Ibid.
13 Ross, *Whaling and Eskimos*, 37.

the Hudson's Bay Company established trading posts north of the tree line in the eastern Arctic.[14]

When a wintering vessel departed at last for her home port in the autumn the profitable economic relationship that had evolved with the native people had to be suspended for a period of ten months or so until the ship could return to Hudson Bay. Some captains, therefore, persuaded Eskimos to collect whalebone, oil, ivory, and skins during the winter and spring, to store them until the ship could return from home port, and then deliver them on board and receive payment in kind. A further dimension was added when whaling firms established some year-round stations on the whaling grounds, manned by a few whalemen or traders. Shore whaling stations operated briefly at Spicer Island in Hudson Strait, at Cape Low on Southampton Island, and at Wager Bay on Roes Welcome Sound in the late nineteenth and early twentieth centuries. In addition a small Scottish vessel, the *Ernest William*, spent several years in Repulse Bay and Lyon Inlet after 1903, functioning as a whaling station and trading post.

From the initial pursuit of one species Hudson Bay whaling interests had become economically omniverous, and by the time of George Comer's voyage of 1903–5 few animals were not of some commercial value. Although the number of vessels entering the Bay and their annual catch of whales had declined, the broad ecological impact of whaling and its socio-economic influence among the Eskimos had increased.

THE ESKIMO

The people we call Eskimos call themselves Inuit ('the people'). Although in earlier times they were never organized in political associations like the Indian tribes they did form groups which occupied specific territories, each group taking for itself the name of the region in which it lived or some dominant feature of the region or its inhabitants.[15] Thus, the Iglu-

14 Wolstenholme post was built in 1909, Chesterfield Inlet in 1911.
15 The word 'tribe' has sometimes been used for Eskimo groups such as the Aivilingmiut, for example, in Franz Boas, *The Central Eskimo* (Washington: Smithsonian Institution, Sixth Annual Report of the Bureau of American Ethnology 1888), 445, and Kaj Birket-Smith, *The Eskimos* (London: Methuen 1959), 144, 233. Other anthropologists have employed the term 'band,' reserving 'tribe' for broader associations of several bands, as in David Damas, 'The Eskimo' in *Science, History and Hudson Bay*, ed C.S. Beals and D.A. Shenstone (Ottawa: Queen's Printer 1968), 1: 146. To Lee Guemple the basic groups are 'regional bands' ('The Institutional Flexibility of Inuit Social Life,' in *Inuit Land Use and Occupancy Project*, ed Milton M.R. Freeman [Ottawa: Indian and North-

lingmiut were the people of the Iglulik region, the Netsilingmiut the people of the seal ('netsik'), and the Aivilingmiut the people of the walrus ('aivik'). The members of a group were unlikely to reside at the same settlement, but they may have come together in particular localities at certain times of the year to co-operate in hunting migrating caribou at river crossings or ringed seals at breathing holes in the winter ice. Dispersion was the usual characteristic of their population distribution but, even so, as families and small bands moved in accustomed seasonal patterns over vast expanses of sea ice and tundra seeking animal food, they possessed not only a territorial consciousness but also a sense of belonging with other families and bands with whom they shared kinship ties and common characteristics of culture and ecology.

Boundaries between groups were not marked but they were known. Each group kept primarily to itself but the relationship between adjacent peoples could vary greatly, according to the degree of similarity or difference in their cultures, the presence or absence of vital resources (such as wood, soapstone, or metal) in their respective territories, and the history of harmony or feuding between them. In northwestern Hudson Bay the Aivilingmiut had peaceful connections with the Iglulingmiut of Foxe Basin and the Tununermiut of the Pond Inlet area, and the cultural similarities among the three groups encouraged some anthropologists to consider them as part of a larger association called the Iglulik Eskimos.[16] The Aivilingmiut had less favourable – even hostile – relationships with groups to the west, however, including the Sinimiut and the Netsilingmiut, whose cultures were somewhat different. South of the Aivilingmiut were the Qaernermiut and other groups belonging to a broader community called the Caribou Eskimos. Their adaptation to life on the barren grounds away from the coast included a strong reliance upon caribou and freshwater fish and set them apart from the Aivilingmiut and other Iglulik Eskimo groups farther north who were primarily coastal-dwelling sea mammal hunters.

When whaling began in 1860 the western regions of Hudson Bay and Foxe Basin were occupied by several groups of Caribou Eskimos between the tree line and the Baker Lake – Chesterfield Inlet area, and two groups of the Iglulik Eskimos farther north. Only the central part of the Hudson

---

ern Affairs 1976], 2: 182). Thomas C. Correll calls them 'demes' or 'miut-groups' ('Language and Location in Traditional Societies,' in *Inuit Land Use*, ed Freeman 2: 173).

16 Therkel Mathiassen, *Material Culture of the Iglulik Eskimos* (Copenhagen: Gyldendal, Report of the Fifth Thule Expedition 1921–24, vol 6, no 1), 1.

Bay – Foxe Basin coast was to be affected directly by whaling. The four principal wintering localities used by the whalemen (Marble Island, Depot Island, Cape Fullerton, and Repulse Bay) were located on the coastal fringe of the territories occupied by the Qaernermiut (one of the northern-most groups of Caribou Eskimos) around Chesterfield Inlet, and the Aivi-lingmiut (the southernmost group of Iglulik Eskimos) in the vicinity of Repulse Bay. The Eskimos of both groups came to play a very important role in whaling activities, and were themselves profoundly affected, as Comer's journal reveals. Equally interesting, however, is the fact that more remote groups, including the Iglulingmiut to the north, the Sinimiut and Netsilingmiut to the west, and the Harvaqtormiut and Hauneqtormiut to the south, were also influenced by whaling. Members of these groups often travelled to the whaling region to trade with wintering ships, and some took up residence. Another group, the Sadlermiut of Southampton Island, attempted to remain outside the influence of whaling but suc-cumbed to introduced disease in 1902.

Most of the names used by Comer for Eskimo groups are the names used by the natives themselves, spelled as he thought best. Most can readily be compared to modern nomenclature, notwithstanding the great diversity of approaches to spelling Eskimo words. The name Kenepetu, however, commonly used through the late nineteenth and early twentieth centuries by whalemen, explorers, and even anthropologists, was a mis-nomer. It is said to have originated when a partly deaf woman replied to the imperfectly heard question of a whaleman concerning her tribal affil-iation, 'It is wet.'[17] The correspondence between the names used in Com-er's journal of 1903–5 and accepted terminology is shown in Table 1.

Before whaling began in 1860 the Eskimo groups inhabiting north-western Hudson Bay and Foxe Basin had been affected very little by direct contact with European culture. They were far beyond the limits of Mo-ravian mission colonization on the Labrador coast, well out of range of the two or three supply vessels sailing annually from Britain to Hudson's

17 Christian Leden, *Uber Kiwatins Eisfelder: Drei Jahre Unter Kanadischen Eskimos* (Leipzig: Brockhaus 1927), 179. Leden's explanation of the derivation of the term dif-fers somewhat from those published earlier. Boas simply conceded that 'the name Kinipetu is said to be derived from Ki' nipoq ('it is wet'), while their proper name is *Kiaknukmiut*' ('The Eskimo of Baffin Land and Hudson Bay,' 6). Another author, apparently obtaining his information from George Comer (as Boas probably did) wrote, 'The people here are called by the rest, Kinnepetu, which may be Englished [sic] 'Damp Place People'' (T.T. Waterman, 'Hudson Bay Eskimos,' *Anthropological Papers of the American Museum of Natural History*, vol 4 [1910], 300).

Table 1. ESKIMO GROUPS

Comer's Names	Usual Destination	Region	Tribal Affiliation
Iglulic	Iglulingmiut	Northern Foxe Basin	Iglulik Eskimos
Iwilic (Iwillic)	Aivilingmiut	Repulse Bay and Wager Bay	
Netchilic (Nectchilic Nectchillic)	Netsilingmiut	Boothia Peninsula King William Island Black River estuary	Netsilik Eskimos
Southampton Island people	Sadlermiut	Southampton	–
Kiackennuckmiut (Kiackennuck) Kenepetu (Kinipetu)	Qaernermiut (Kiaknukmiut)	Chesterfield Inlet	Caribou Eskimos
Show-vock-tow-miut (As-shock-miut)	Saquaqturmiut? Asiarmiut?	Baker Lake and south of Chesterfield Inlet	

NOTE: For populations of the major groups compiled in sex and age categories see appendix I.

Bay Company posts in southern Hudson Bay and James Bay, and too far north to make regular visits to the Company's northernmost post at Fort Churchill. The attempt by the Company during the eighteenth century to extend trading relationships northward with summer sloop voyages from Fort Churchill did not reach beyond Rankin Inlet, so that the people farther north, who later felt the full impact of commercial whaling, were not directly affected.

Explorers had visited the west coast of Hudson Bay from time to time, beginning in the seventeenth century, but the influence of these expeditions was not great. Ship-borne explorers seeking a northwest passage to the Pacific tended to ignore or intentionally avoid the native people, and they seldom stayed long in one locality. Their infrequent (usually unintentional) winterings occurred mainly in James Bay and the southern margins of Hudson Bay. The expedition commanded by James Knight in 1819–21 did winter at Marble Island, but what associations were established with the Eskimos of the mainland before all the men of the party died of starvation, scurvy, and exposure, cannot be known. Before the whaling period, therefore, Eskimo – white contacts were few and far between and did not result in any significant modification of the aboriginal culture.

Eskimos of the whaling region received a few items of European manufacture through occasional contact with transient explorers, and a few individuals from southern parts of the whaling zone may have journeyed to the trading post at Fort Churchill to barter for articles. But the distance to the post, the presence near Fort Churchill of Padlimiut middlemen anxious to preserve their own favourable trading position, and the absence of umiaks (large, open, skin 'luggage' or 'family' boats) on the west coast of Hudson Bay during the nineteenth century discouraged regular connections with the Company post. Travelling along the coast in 1846–7 John Rae observed that the economic importance of the Fort Churchill post steadily declined as he went north.[18] Beyond Depot Island few if any Eskimos had ever visited it.

None the less, Eskimos of northwestern Hudson Bay and Foxe Basin possessed a modest assortment of alien goods. In 1820–2 the officers of Parry's second expedition in search of a northwest passage observed iron blades, copper kettles, files, knives, needles, beads, and other articles among the natives of Winter Island and Iglulik, 650 and 900 miles respectively from Fort Churchill, the closest trading post. Through a series of transactions, Parry concluded, the goods had passed from group to group, reaching the remote Iglulingmiut well ahead of the first face-to-face encounter with Europeans. Although the geographical range of this pre-contact inter-tribal exchange was impressive, the volume of trade must have been small and the variety of goods exchanged was certainly limited. The acquisition of some metal tools, containers, and materials increased the effectiveness of traditional operations and processes but did not fundamentally alter the Eskimo culture.

The initiation of commercial whaling in 1860 resulted in the first regular and prolonged contacts between the native inhabitants of northwestern Hudson Bay and Euro-Americans, and provided the first opportunities for the Eskimos to obtain a variety of outside articles directly through trade and employment. For the next half-century, as a result of the Eskimos' desire for material goods on one hand and the whalemen's need for Eskimo help on the other, there was relatively continuous interaction between Eskimos and whites in the region between Marble Island and Lyon Inlet. To overwinter successfully, the whalemen required the assistance of native seamstresses to make up fur clothing and native hunters to bring in fresh meat – mainly caribou – to supplement the shipboard diet of preserved food and to prevent scurvy. During the floe whaling of May and June they needed Eskimo men with sleds and dog teams to

18 John Rae, *Narrative of an Expedition to the Shores of the Arctic Sea in 1846 and 1847* (London: Boone 1850), 182, 184.

transport whaleboats, equipment, and supplies to the floe edge and carry bone and blubber back to the ship after each whale kill. As money played no part in the Eskimo economy the whaling masters paid their 'ship's natives' in material goods such as tools, weapons, whaleboats, and whaling gear.

It quickly became standard practice for captains to take an assortment of trade goods north on each voyage for the purpose of paying off ship's natives and trading with other Eskimos for furs and ivory. By the time of Comer's voyage the trading outfits of whaling ships were impressively large and varied. The *Era* carried two dozen rifles, more than 13,000 primed shells, 85,000 primers, 10,000 percussion caps, 1,200 fish hooks, 11,000 needles, more than 700 yards of calico, and almost 200 knives, as well as hatchets, saws, files, steels, gimlets, awls, forks, spoons, scissors, beads, and many other articles (see appendix c).

Of all the items introduced by whalers guns and whaleboats were undeniably the most important. Unlike metal pots, needles, and knives, which had more primitive counterparts of stone and ivory in the traditional culture, muskets and rifles were completely alien to Eskimo life. They were highly prized because they provided much more striking power, range, and accuracy than bows or harpoons. In winter, when lack of cover and hard, crunchy snow made it difficult for a hunter to get close enough to solitary or scattered caribou to use a bow, a rifle enabled him to shoot from a greater distance with a good chance of success. Guns therefore improved the chances of surviving inland in winter, and probably encouraged longer hunting trips and traplines. In late winter a hunter normally had to approach basking seals by sliding stealthily along the sea ice imitating another seal, until he could get close enough to run forward and throw his harpoon, but the range of the rifle reduced the stalking time and the risk of being spotted by the prey. Shooting seals from the floe edge or from a boat was easier with a rifle than a harpoon because of the greater accuracy of the gun and velocity of the bullet. Polar bear and musk-oxen could be hunted with far less danger. The adoption of firearms led to a number of changes in hunting methods and generally improved the chances of success.

Certain disadvantages offset to some extent the power and accuracy of the rifle, however. Whereas the harpoon and line facilitated the retrieval of dead or wounded sea mammals, the gun was solely a killing instrument, and many animals sank or were swept away by currents. Furthermore, firearms required a continuing supply of caps or flints, and balls or bullets, binding the hunter economically to foreign manufacturers and suppliers.

Nevertheless, guns were always in great demand during the whaling period. By 1903 every hunter among the Aivilingmiut and Qaernermiut probably possessed at least one gun. The traditional bow was by then obsolete.

The acquisition of whaleboats was of special significance to the Eskimos of western Hudson Bay because they had not previously possessed umiaks or any other large boats in the nineteenth century. They had kayaks for coastal and inland use, but the kayak was usually a single-occupant hunting craft of little use in transporting bulky cargoes or large numbers of people. Kayaks were sometimes used for hunting walrus and even bowhead whales, employing harpoons with floats attached, but they were low, light, and easily holed, and hunters had to stop paddling in order to throw their harpoons. Second-hand whaleboats introduced as wages and in trade to the Aivilingmiut and Qaernermiut filled an important gap in their material culture, providing a large, open, seaworthy, durable, general-purpose boat for coastal travel and sea mammal hunting. Whaleboats were observed among the Aivilingmiut as early as 1864, only four years after whaling began.[19] By 1903 there were at least twenty between Chesterfield Inlet and Repulse Bay – roughly one boat for every two or three families.[20] With these boats the Eskimos improved their chances of obtaining employment with the whalers, and extended the range of their own hunting trips and travel. Around the turn of the century a Baker Lake man sailed his whaleboat to Repulse Bay and back – a total distance of more than 700 miles – merely to find out where the whaling ships were going to pass the winter.[21]

Guns and whaleboats had a noticeable impact on Eskimo hunting and travelling, but every article of Euro-American manufacture, however small, played a role in modifying Eskimo life. Some items, which had no equivalent in the traditional culture, by making possible completely new activities and processes, enriched the material culture. Telescopes increased the range of hunters' vision. Fiddles and accordions enlarged the dimensions of music. Cribbage boards, dominoes, and footballs added variety to recreation. Flour, molasses, coffee, and hard tack brought culinary delights (and dental caries) of a new order. Other introduced articles performed the same functions as implements already present in the ab-

---

19 Charles Francis Hall, *Narrative of the Second Expedition Made by Charles F. Hall ... 1864–69*, ed J.E. Nourse (Washington: Government Printing Office 1879), 63.
20 Ross, *Whaling and Eskimos*, 95.
21 David Theophilus Hanbury, *Sport and Travel in the Northland of Canada* (London: Arnold 1904), 60.

original culture but had certain important advantages, such as greater durability, strength, lightness, or the ability to hold a sharp edge or point. Metal pots, kettles, and knife blades, canvas tents, rod iron harpoon shafts, steel traps, needles, and sunglasses were more effective or more convenient in the snow house, on the trapline, and in the hunt, but the adoption of such items usually resulted in the atrophy of traditional counterparts, contributing to the decline of native skills.

To obtain maximum benefits from employment the Eskimos had to be at the ships when needed. They therefore reduced their traditional mobility and dispersion, taking up semi-permanent residence at winter harbours occupied by whaleships. The location of the whalers came to govern the winter distribution of the Aivilingmiut and Qaernermiut, and even some members of more distant groups. If it became known that ships would be present in both Repulse Bay and Marble Island the Aivilingmiut would probably remain in Repulse Bay – their home territory – and the Quaernermiut would probably settle opposite Marble Island. But if the ships concentrated in one harbour then both groups would likely relocate there for the winter. Whaling thus brought about a centralization of Eskimo population in winter, a 'shifting centralization' in which the location of the large winter settlements often varied from one year to the next, depending on the harbours selected by the whaling masters.

The attraction of Eskimos to winter harbours eroded barriers of hostility and suspicion, encouraged the transfer of individuals from one group to another, and led to remarkable alterations in the pattern of Eskimo distribution. The Qaernermiut enlarged their territory northward beyond Chesterfield Inlet. The Aivilingmiut expanded southward from Repulse Bay along the mainland coast and in 1908 colonized Southampton Island, whose native inhabitants, the Sadlermiut, had by then been eliminated by disease. As the Aivilingmiut spread south and east a number of Netsilingmiut took up residence around Repulse Bay, previously the focal point of Aivilingmiut existence.

Eskimos employed by the ships frequently associated with whalemen, both on the job and in the sporting and recreational activities that took place in the winter months. Eskimo men and women often joined with sailors in football games, dances, and concerts; the informal intermingling was a learning process for both and contributed to mutual understanding. But the path towards acculturation is seldom smooth. With the material and social benefits of close association at the winter harbours came unfamiliar diseases which on occasion struck down the native people with dreadful effectiveness, illicit sexual unions which resulted in a number of half-breed children with absentee fathers, and some exposure to the harmful effects of alcohol.

Long before 1903 contact with whalers had become a normal part of the annual cycle of the Eskimo groups of northwestern Hudson Bay, who had come to depend upon whaling captains for employment and the material goods obtained as wages. Despite the declining number of vessels sailing north each year, relationships with the whalemen had remained remarkably continuous. Harry, Comer's head native in 1903–5, is known to have worked for whalers at least eight winters between 1893 and 1905. His brother Ben, who had saved Comer from drowning in February 1894, also worked eight or more winters for whalers until his death on 2 June 1905. Melichi, another man employed by Comer in 1903–5, had been a regular ship's native long before 1878 and carried on until his death about 1910, by which time he had probably been employed by whalers most of his adult life.

Some whaling captains were fair in their dealings with the Eskimos and considerate of their welfare. It was not unusual for them to supply medical help for sickness or accident, to distribute food when hunts were unsuccessful, to invite natives to social events, and to exercise a measure of supervision and control over the potentially harmful activities of their own crews. In general Hudson Bay masters were successful in preventing excessive use of alcohol on board ship and among the Eskimos; indeed, many commanded vessels designated by their owners as temperance ships. Few whaling captains, however, could have felt as much concern for the native people as George Comer. To the systematized procedures and relationships of employment and trade he added a distinctly personal dimension, treating the people as individuals and establishing close friendships among them. He respected their culture, worked hard to record it, and worried about the insecurity of traditional life on the land. To the Aivilingmiut and many others Comer was a friend. His fellow-whaleman simply regarded him as 'native crazy.'

GEORGE COMER

George Comer was allegedly born in the city of Quebec, Canada, on 22 April 1858, of English parents Thomas and Johanna Comer.[22] His father

22 Quebec City is frequently cited in newspaper articles, obituaries, and biographical notes as Comer's birthplace (see, for example, Barnard L. Colby, 'George Comer,' *New London Whaling Captains* [Mystic, Conn: Marine Historical Association 1936], 223). His descendants today agree that he was born in Quebec. The official historian

was lost at sea three years later and following his mother's subsequent move to the United States young George, then only six, was placed in an orphanage in Hartford, Connecticut. When he was ten years old, he left to work on the farm of one William H. Ayres in East Haddam, a village some twenty-five miles south of Hartford, and evidently remained several years. We do not know what influences or events were responsible for George's decision at the age of seventeen to enter the whaling trade, but in 1875 he made his way to the port of New London and signed onto the bark *Nile*, bound for arctic regions under the command of Captain J.O. Spicer.

George kept a diary on this voyage, a practice which he continued during the rest of his career.[23] The little notebook, its entries written in pencil and only slightly faded a century later, records the principal events of the trip in the briefest of terms. At the outset of the voyage there was a period of adjustment to the ship, her crew, and the sea. Early entries reflect the misery of the transformation from landlubber to seaman, and perhaps from boy to man: 'I have been homesick every day, was sea sick the first too [sic] days and lost my hat overboard;' 'I guess you are having huckleberry and milk for supper to night. Don't I wish that I was home with [you] to night.' In a few months, however, there was so much to occupy the interest of a young man on his first voyage into the Arctic that his thoughts strayed less frequently to the comforts and pleasures of home. He wrote of encounters with other vessels, 'gams' among their captains (which on one occasion resulted in Captain Spicer returning 'blind drunk'), sightings of icebergs, bird-shooting excursions, a fight between the Captain and an insubordinate seaman, eating whale meat for supper, and trading with the Eskimos of Baffin Island for skin clothes, boots, mittens, and slippers. There were some moments of excitement and danger. Severe gales carried away two whaleboats. A sudden leak kept the men pumping half an hour out of every two hours. A huge whale smashed the captain's boat, rose so close to George's boat 'that I could put my hand on his head

---

of the Canadian government expedition on the steamer *Arctic*, who had plenty of opportunity to talk with Comer during the winter of 1904-5, described him as 'un Quebecquois [sic]' (Fabien Vanasse, 'Relation sommaire du voyage de l'*Arctic* à la Baie d'Hudson 1904-1905,' 30). Nevertheless, inquiries to the registre de la popula-tion, the palais de justice (archives civiles), and the centre d'archives de la capitale (service de la généalogie), in the city of Quebec, have so far failed to reveal any record of his birth.

23 George Comer, Manuscript Journal on Board the *Nile* 1875-6, George Comer Papers, East Haddam, Conn.

but I did not want to,' received no fewer than eleven explosive projectiles, and yet somehow escaped. The youthful George Comer appears to have taken these adventures in stride, along with the discomfort and hardship that were the inevitable companions of the arctic whaleman, devoting no more than a few pencilled lines to happenings which others would surely have regarded as extraordinary.

Following the arctic voyage on board the *Nile* Comer worked in the American coastal trade for three years, and in 1879 sailed for the bottom of the world on the *Mary E. Higgins* of New London. The small schooner of ninety-eight tons cruised in the treacherous waters near Magellan Strait and Cape Horn and returned in 1880 with a thousand sealskins, then departed again for antarctic seas in the same year, remaining out until April 1882 and securing over 2,000 skins.[24] In the following year Comer shipped on board the *Charles Colgate* of New London bound for the southern Indian Ocean to procure the oil of sea elephants (elephant seals). The schooner sailed to Desolation Island, now called Kerguelen Island, a forbidding volcanic mass whose rocky coasts supported vast colonies of the world's largest species of seal. Sometimes exceeding twenty feet in length and weighing as much as three tons, the huge animals were generously provided with blubber, and their habit of congregating in shore rookeries, combined with their awkwardness on land, made them easy victims. The hunters would simply locate a colony, cut the animals off from the sea, and beat them to death with five-foot clubs. Blubber would be sliced off to be tryed out, and the rest of the formidable carcasses left on the rocks to rot. The *Charles Colgate* arrived back in New London in the spring of 1884 with 1,100 barrels of oil.[25] In 1885 Comer joined the seventy-ton schooner *Express* of Stonington as mate for a sealing voyage to South Georgia. After the tiny vessel returned in April 1886 with a small quantity of oil he set out again for Desolation Island, this time on the *Francis Allyn* of New Bedford, which stayed out for almost two years and secured 390 barrels of oil.[26]

The antarctic voyages of 1879–89 were the foundation of Comer's sub-

24 George Comer, 'Voyages That I Have Made,' in Account Book of the *A.T. Gifford* 1907–9, G.W. Blunt White Library, Mystic Seaport, Mystic, Conn, end pages.
25 Reginald B. Hegarty, *Returns of Whaling Vessels Sailing from American Ports: A Continuation of Alexander Starbuck's 'History of the American Whale Fishery' 1876–1928* (New Bedford, Mass: Old Dartmouth Historical Society 1959), 15.
26 Ibid. 22.

sequent career in the arctic regions. There is no tougher proving ground for sailors than the seas surrounding the ice-clad continent of Antarctica. The 'roaring forties' and adjacent polar waters are famous for their strong gales, immense breaking seas, near-freezing water temperatures, concentrations of pack ice and icebergs, and absence of shelter – a distinctly forbidding combination of environmental conditions. A voyage through antarctic seas is now regarded by yachtsmen as the ultimate challenge notwithstanding the advantages of steel hulls, auxiliary engines, radar, echo sounders, direction finders, radios, and self-steering.[27] For nineteenth-century sealers (who did not possess as much as a radio) it was simply an accepted procedure for obtaining seal oil and skins. When George Comer returned in 1889 to the whaling grounds of the Canadian Arctic, it was not as a young, homesick greenhand, but as a tough, experienced seaman, competent in the working of small sailing ships among the hazards of severe gales, cold weather, fogs and sea ice. The arctic whale fishery was inheriting a capable mariner and a confident leader.

George Comer's return to arctic whaling, fourteen years after his initial experience on the *Nile*, was inaugurated with three one-season voyages to Baffin Island as first mate of the *Era*, serving again under his first mentor, Captain J.O. Spicer. The *Era* supplied the whaling stations operated by the New London firm of C.A. Williams. These land bases were usually staffed by two white men, who in turn employed a number of Eskimo families in whaling, hunting, and trapping. At each station, the *Era* was to put ashore the food, provisions, trade goods, and coal required for the forthcoming year, take off the produce secured since the last visit, and carry out any necessary exchange of personnel.

The *Era* departed from New London on 10 July 1889, entered Hudson Strait on 2 August, and arrived at the Spicer Island station on the fifth.[28] A few days sufficed for the unloading and loading of goods, and the vessel then sailed out of the Strait, turned northward across the mouth of Frobisher Bay to visit the station at Cape Haven, and continued to Blacklead Island in Cumberland Sound. The itineraries of 1890 and 1891 were essentially similar to that of 1889, and in these three voyages the *Era* collected from the Baffin Island stations a total of 523 barrels of oil, 6,689

27 See David Lewis, *Ice Bird: The First Single-Handed Voyage to Antarctica* (New York: W.W. Norton 1975).
28 George Comer, Manuscript Journal on Board the *Era* 1889, George Comer Papers, East Haddam, Conn.

pounds of bone, several hundred pounds of ivory, and the skins of at least 100 foxes, 39 bears, and a few wolves.[29]

Compared to his earlier experiences in the Southern Ocean the arctic voyages of 1889–91 must have seemed somewhat dull and limiting to George Comer. The *Era*'s primary task was to supply the shore stations, so there was little time to hunt whales, and the brief duration of the visits to the stations, made necessary by the uncertainties of sea ice and weather, provided little opportunity to indulge his curiosity about the natural world.[30] During a few idle moments at Cape Haven in August 1890 he was able to collect some rocks which appeared to contain high concentrations of iron, but there was no time to do more. Eskimos were sometimes at the harbours when the ship arrived, but again limited time made it impossible to develop proficiency in their language or learn much about them. To see fascinating objects of inquiry but be denied the opportunity to examine them thoroughly must have been deeply frustrating.

When the *Era* departed again for the arctic stations in 1892 Comer was not on board, but in the following year he joined the New London bark *Canton* under Captain Elnathan B. Fisher for a wintering cruise into Hudson Bay, a voyage which was to give greater rein to his scientific interests and establish a pattern of activities for the next two decades of his life. The vessel departed from New Bedford on 24 June 1893 and was in sight of Resolution Island by 26 July.[31] Hudson Strait ice was heavy and it was not until 14 August that the *Canton* was abreast of Nottingham Island with clear sailing ahead through Fisher Strait to the whaling grounds.

29 The oil and bone returns are from Hegarty, *Continuation of Starbuck*, 24–6. The fur returns are from daily entries in George Comer's journals on board the *Era* in 1889, 1890, and 1891, George Comer Papers, East Haddam, Conn, and probably understate the amounts obtained.

30 Comer had returned from his 1887–9 antarctic voyage on the *Francis Allyn* with four skins and two live specimens of an unknown species of flightless bird from Gough Island, subsequently named in his honour *Porphyriornis comeri* (J.A. Allen, 'Description of a New Gallinule, from Gough Island,' *Bulletin of the American Museum of Natural History* 4 [1892]: 57–8). According to ornithologist Robert Cushman Murphy, Comer collected a number of other antarctic natural history specimens for Yale University and the American Museum of Natural History, and was the first to obtain a photograph of a King Penguin incubating an egg (Robert Cushman Murphy, 'Remarks at the dedication of the Captain George Comer Memorial ...' Typescript, George Comer Papers, East Haddam, Conn, 1, 2).

31 The synopsis of this voyage is based on George Comer, Manuscript Journal on Board the *Canton* 1893–4, John Hay Library, Brown University, Providence, RI and Edgar W. Crapo, Manuscript Logbook on Board the *Canton* 1893–4, G.W. Blunt White Library, Mystic Seaport, Mystic, Conn.

One whale was killed on 16 September by Comer's whaleboat crew, and on the twenty-ninth Captain Fisher took the ship into winter quarters at the inner harbour of Depot Island, in company with the bark *A.R. Tucker.* For the next nine-and-a-half months the two ships remained frozen into the ice while their crews slept, ate, amused themselves with football, concerts, and dances, and enjoyed intimate relationships with Eskimo women. Floe whaling commenced in late April and in June the sailors were put to work sawing a narrow channel through ice several feet thick from the ships to the open water a mile-and-a-half away, an arduous task which continued around the clock. After taking several whales in the second summer the *Canton* arrived home on 16 October 1894 with 200 barrels of oil and 6,000 pounds of bone.[32]

Comer returned to Hudson Bay in the summer of 1895, sailing again on the *Era,* the small schooner that had carried him to Cumberland Sound on three previous occasions. On this cruise, his first as master, he elected to winter among islands off Cape Fullerton, a locality to which he was to return on many subsequent voyages.[33] The ship secured one whale in the first summer and two in the second, returning in October 1896 with 240 barrels of oil and 6,000 pounds of bone.[34] In 1897 the *Era* left again for Hudson Bay on a two-year cruise, the first time since the discovery expedition under Captain W.E. Parry, RN, in 1821-3 that a crew passed two successive winters on board ship in the Hudson Bay region.[35] With the help of native boat crews Comer secured fifteen whales and delivered to New Bedford wharves on his return in September 1899 a total of 285 barrels of oil and 18,000 pounds of bone.[36] In the following summer he headed north again on the *Era.*[37] The ship wintered twice at Fullerton Harbour, as it had on the previous cruise, employed a number of Eskimos, and came back with 130 barrels of oil and 7,000 pounds of bone, as well as over 800 skins of musk-ox, polar bear, arctic fox, wolverine, and wolf.[38]

32 Hegarty, *Continuation of Starbuck,* 29.
33 The details of this voyage are taken from a mate's manuscript journal written on board the *Era* 1895-6, Old Dartmouth Historical Society, New Bedford, Mass.
34 Hegarty, *Continuation of Starbuck,* 31.
35 The details of this voyage are taken from George Comer, Manuscript Journal on Board the *Era* 1897-9, G.W. Blunt White Library, Mystic Seaport, Mystic, Conn.
36 Hegarty, *Continuation of Starbuck,* 32.
37 Details of this voyage are taken from George Comer, Manuscript Journal on Board the *Era* 1900-2, G.W. Blunt White Library, Mystic Seaport, Mystic, Conn.
38 Hegarty, *Continuation of Starbuck,* 35.

All four Hudson Bay voyages made by George Comer from 1893 to 1902 involved wintering in the Arctic, and two included two successive winters there. In the long, relatively idle months of winter he was able at last to pursue his interests in natural phenomena and Eskimo culture. Not only was more time available for his own pursuits, but also from 1895 on he sailed as master rather than mate, with the authority to make decisions on the choice of winter harbours, the hiring of Eskimos, the timing and extent of spring whaleboat cruises, and other matters, decisions which ultimately affected his own ethnographic and scientific interests. He collected plants and bird skins for professors at Yale. He learned what he could about Eskimo burial customs, spiritual séances, and taboos. He took photographs of various aspects of Eskimo life, examined ancient graves, compiled population figures for the principal groups (see appendix I), recorded the names, ages, heights, and weights of Eskimos of both sexes, and made plaster casts of a number of heads, hands, and feet.

His thoroughness in collecting artifacts of prehistoric and contemporary culture, as well as natural history specimens, established him as a dependable field man for scientific institutions, and in subsequent years more than one authority acknowledged his work. When the *Era* departed in the summer of 1903 for Hudson Bay, Comer had been commissioned to obtain a 'full' ethnographic collection for the Museum für Völkerkunde in Berlin, and to secure for the American Museum of Natural History in New York some clothing, charms, and amulets from the Netsilingmiut, clothing from the Sadlermiut, plaster casts of heads, hands, and feet, recordings of songs, and notes on customs.[39]

THE WHALING JOURNAL

Generally speaking, whalemen were not much inclined to record their experiences. Out of more than 2,500 nineteenth-century voyages to Davis Strait and Baffin Bay there are only a dozen or so published full-length narratives. With the exception of Ferguson's *Arctic Harpooner*, an account of a cruise on the *Abbie Bradford* in 1878, there are no published first-hand records of whaling in Hudson Bay. The story of arctic whaling lies hidden in the unpublished ships' logbooks and private journals that have survived the ravages of time. Such documents have in recent years ac-

39 Franz Boas to Comer, 15 April, 19 May, 5 June 1903, and Boas to Karl von den Steinen, 1 May 1903, American Museum of Natural History, New York; J.A. Allen to Comer, 18 June 1903, George Comer Papers, East Haddam, Conn.

quired extraordinary value as antiquities, especially when illustrated by sketches and by whale stamps denoting the kills, and those few remaining in private hands are either jealously guarded as treasures of the past or have entered the collector's world of exchange, a milieu in which they are relatively inaccessible to the scholar. Fortunately a number of historical societies and museums have amassed impressive collections of whaling journals and logbooks through the years, before escalating prices put them beyond the reach of all but the most richly endowed institutions, and they can be used to piece together the development of whaling in the arctic regions and assess its impact on the native peoples. Not all regions are well documented, however. Only 6 per cent of the American and British voyages to the Davis Strait whaling grounds in the nineteenth century are represented by logbooks or journals. In Hudson Bay, on the other hand, documents of this sort exist for almost half the voyages, and the history of whaling can be reconstructed with greater reliability.

Journals and logbooks vary enormously in content, accuracy, and readability. The least interesting and least informative are logbooks in which a semi-literate mate of limited imagination has entered day after day brief notes such as 'Temperature 46 all well so ends this day,' or even terser, 'All well so ends.' Some private journals, on the other hand, written perhaps by a master, a surgeon, or an educated young man shipping before the mast for the sake of adventure, record in detail not only events of exceptional interest but also the more routine aspects of shipboard life, the operation of the vessel, whaling techniques, and other matters which, although familiar enough to the whalemen themselves, were not necessarily well known to the general public. Keeping a written daily record of events, as most of us know from experience, requires a good deal of self-discipline, not to mention a certain talent for observation, reflection, and self-expression, and it is no surprise that few of the seamen and officers on board whaling ships possessed these qualities in sufficient measure to compile a continuous and reasonably complete record during an entire cruise. George Comer was an exception. He appears to have kept a diary on every one of his voyages, and with few exceptions he made entries every day, even during the floe whaling of April and May, when the men cruised among ice floes, constantly cold and damp, eating and sleeping in the small whaleboats. He knew that a record of the events within his own experience would be valuable not merely as a memento but as a historical document as well. The renowned arctic explorer and author Vilhjalmur Stefansson once chided him for lending several of his whaling journals to someone who had not returned them in over a year. 'Do get after him and see what he is doing,' he urged. 'They are priceless historical and ethnographic material. Who is the man? If he is a New Yorker I could perhaps hustle him up for you.' [40] In fact, Comer knew that his journals could prove useful in the future and he must have taken care to impress the fact upon others, including his own family, for most of the documents have survived and are now in safe hands.

In later life George Comer used to make periodic visits to deteriorating cemeteries around East Haddam, tidying them, repairing broken monuments, and cleaning the inscriptions. It was a responsibility he took upon himself because he was 'full of reverence for historical records.'[41] He respected those who had gone before and any visible evidence of their achievements. He felt the same about the Eskimos, treasuring their legends and myths, their stories of the past, the artifacts that revealed earlier cultures. He recorded their customs, their numbers, and the changes in their geographical distribution over the years. This historical and ethnographic dimension gives added interest and value to his arctic journals.

George Comer's narrative of 1903–5 is written in pen in two hard cover notebooks, each $8^1/_4$ by $13^1/_2$ inches, with lined, numbered pages. The first notebook, from 29 June 1903 to 31 October 1904, contains 192 pages of observations and the second, from 1 November 1904 to homecoming on 15 October 1905, includes 146 pages of writing. The journal entries thus run to a total of 338 pages and about 55,000 words. At the back of each notebook is miscellaneous information relating to the operation of the vessel on the voyage, including itemized inventories, at specific dates, of ship's stores, slops, whaling gear, trade goods, and skins secured from the Eskimos, records of whale kills and of ice thickness at the winter harbour, and details on a variety of other topics, such as skin boots made for the sailors by native women, large knives given to certain native men, and a wage agreement with ship's native 'Smiley.' Other supplementary items are to be found written or pasted into journal pages, or inserted loose here and there. There is a photograph of the *Era*, a watercolour sketch of the vessel, a list of whales sighted on a previous voyage in 1895–6 and one of whales struck in 1897–8, a description of the loss of the *Era* in 1906, and an issue of the *Whalemen's Shipping List and Merchants' Transcript*. Several newspaper clippings are in the preliminary pages of the notebooks, three pertaining to the 1903–5 voyage and one discussing

40 Stefansson to Comer, 18 August 1927, George Comer Papers, East Haddam, Conn.
41 P. LeRoy Harwood, 'P. LeRoy Harwood at Capt.Comer's Memorial Dedication,' Typescript, George Comer Papers, East Haddam, Conn, 5.

arctic exploration, drawing attention to Comer's earlier collection of specimens for the American Museum of Natural History. In all some two dozen items supplement Comer's day-to-day narrative of the 1903–5 voyage to Hudson Bay. Several of the most important are included in this volume as appendixes.

At some date prior to George Comer's death in 1937 this journal was among several borrowed by his friend P. LeRoy Harwood, treasurer of the Mariners Savings Bank of New London, who was greatly interested in the history of whaling. In 1939, following Harwood's death, the journals were presented to Mystic Seaport, Connecticut, and are today part of the archival holdings of the G.W. Blunt White Library there.[42]

In editing Comer's journal I have adopted what the *Harvard Guide to American History* calls the 'modernized method,' the object being 'to make an early document, chronicle, or narrative intelligible to the average reader who is put off by obsolete spelling and erratic punctuation.'[43] To preserve the manuscript's quaint spelling and inconsistent capitalization, punctuation, and paragraph structure would achieve nothing but confusion and misunderstanding. Therefore, in the interest of clarity, comprehension, and consistency, I have corrected spelling according to American usage, followed official nomenclature for Arctic place names,[44] provided punctuation, retained or supplied capitalization where necessary, expanded abbreviations (except obvious ones repeated in daily reports of ship's course, such as wind direction, temperature, pressure, and so on), and standardized the form of numbers, dates, and daily entries. Grammar has been altered only to the extent of correcting a few glaring mistakes in past particples and agreement of number. There have been no changes in sentence structure or sequence, although the provision of punctuation has necessarily involved some degree of judgment about the intended structure and meaning. When words have inadvertently been repeated in the manuscript I have eliminated the repetition. When a word has been carelessly omitted in the original I have supplied it in square brackets. Illegible words or portions of words have been indicated by a dash, the inferred word or letters following in square brackets. For the reader's convenience I have divided the journal into chapters. The document is, after all, a diary,

---

42 The official designation for the Comer manuscript is Coll. 102 (vols 3 and 4). A copy exists on microfilm rolls 90 and 70.

43 Frank Friedel, *Harvard Guide to American History* (Cambridge, Mass: Harvard University Press, Belknap Press 1974), 1: 31–2.

44 *Gazetteer of Canada: Northwest Territories*, provisional ed. (Ottawa: Surveys and Mapping Branch, Department of Energy, Mines and Resources 1971).

written in the Arctic in the uncomfortable, cramped quarters of whaling schooners and whaleboats at odd moments in a busy routine. Had its author decided to submit it for publication he would surely have desired at least the extent of editorial alteration provided here.

In the introduction, appendixes, and notes I have retained the traditional English word 'Eskimo' unless referring to specific groups (for example, Aivilingmiut). The terms 'Inuit' (pl.) and 'Inuk' (s.), used by most of the native people of the Canadian Arctic, present certain grammatical difficulties in English, including number and adjectival use, whereas 'Eskimo' may be used correctly as singular or plural, noun or adjective, and is widely understood. I see nothing derogatory in the term.[45]

There are several reasons why George Comer's 1903–5 journal should be published. First, and most simply, it is a useful addition to the sparse literature of Hudson Bay whaling. Although whaling in this region was overshadowed by the long and intensive fishery in Davis Strait and Baffin Bay it had unique and interesting aspects which should be known. It was dominated by American whalemen rather than by British, the fleet consisted of sailing vessels rather than steamers, and the wintering of ships and crews was much more common. Second, Comer's journal provides a portrait of the last stages of arctic whaling. The demand for baleen in the late nineteenth century had given arctic whalers a new market while the whaling industry was declining swiftly elsewhere but, as the bowhead whales had become more and more scarce, whaling firms had found it increasingly necessary to engage in the fur trade. The economic emphasis in 1903 was, therefore, substantially different from that prevailing in 1878, the time of Ferguson's voyage on the *Abbie Bradford*.[46] Third, the journal reveals a good deal about the social and economic relationships between Eskimos and whalemen. When whaling began in 1860 the native inhabitants of the west coast of Hudson Bay had experienced only occasional contact with white explorers and trading posts, but by 1903 a few hundred Aivilingmiut, Netsilingmiut, and Qaernermiut were heavily dependent upon the whaling ships for the supply of weapons, ammunition, boats, tools, utensils, clothes, and tobacco, which had become integral parts of their lives. Fourth, Comer's journal provides insight into the issue of Canadian

---

45 See Terence Armstrong and Hugh Brody, 'The term "Eskimo," ' *Polar Record*, 19, 119, (1978): 177–90; Albert C. Heinrich, 'Letter to the Editor,' *Arctic*, 33, 1 (1980): 204–5; Michael C.S. Kingsley, 'Use of the words 'Inuk' and 'Inuit,' *Arctic*, 32, 3 (1979): 281.
46 Robert Ferguson, *Arctic Harpooner: A Voyage on the Schooner Abbie Bradford 1878–79*.

sovereignty in the Arctic. Until the time of this voyage American and British whalers had operated in the waters of northern Canada without supervision or restriction, but at last the Canadian government decided to exert authority in the Arctic. It sent an expedition into Hudson Bay in 1903 to regulate the whaling industry and dispatched a second expedition in 1904. Although George Comer could not know it when the *Era* left New Bedford on 30 June 1903, he was destined to spend the next two winters in the company of Canadian government steamers and a detachment of the Royal North-West Mounted Police.

THE SOVEREIGNTY ISSUE

Before 1903 there had been thousands of whaling cruises to the waters of northern Canada, but few if any of the whaling masters had given any consideration to the matter of national sovereignty over the vast regions in which they cruised. They saw the Arctic as an immense area, rich in wildlife resources there for the taking, inhabited by a small and dispersed population of primitive Eskimos, and completely devoid of any political control. In the absence of any instructions to the contrary they felt themselves at liberty to take as many whales as they could, to feed wintering crews upon caribou and musk-oxen of the adjacent coasts, to employ Eskimos for a variety of tasks, and, in return for their goods and services, to introduce guns, boats, clothes, food, and a variety of implements and materials, all without import duties, with no thought of the ecological consequences, and at standards of barter set by themselves which in retrospect appear distinctly unfair to the natives.[47]

At no time did the whalemen entertain the idea that the land, the waters, and the resources might belong to the aboriginal people who had resided there for centuries, and that the commercial exploitation of these regions should be carried out only with the consent of the resident population. Such views, which now appear entirely reasonable and justified and are being put forward with vigour by the native societies of the north, were then non-existent, among the whites at least. Territorial rights were for 'civilized' people, not for nomadic hunters. As for evidence of the

---

47 'The return given the Esquimaux for valuable furs and whalebone is a mere nothing. As an example I may quote that 100 primers for '38 or '44 calibre Winchester rifles are considered a fair exchange for a musk ox robe. The primers cost $1.08 per 1,000 in the United States.' J.D. Moodie, 'Report of Superintendent J.D. Moodie on Service in Hudson Bay ... 1903–4,' 11.

authority of a national government, or even of southern commercial influence, there was none. The hauntingly beautiful arctic landscape had not yet been marred by the discordant shapes and colours of buildings, airstrips, drilling rigs, abandoned vehicles, and empty fuel drums. In the absence of any forceful expression of native rights or any clear demonstration of Canadian sovereignty, it is not at all surprising that the whalemen felt at liberty to act with a free hand in the Canadian Arctic, and to exploit the biological and human resources as they wished. Their ignorance of basic geographical relationships is symbolized by the page heading of a whaling logbook of 1867 in northwestern Hudson Bay which reads, 'Lying in Repulse Bay, Greenland.'[48]

As a matter of fact ownership of the arctic regions of northern Canada was not precisely clear.[49] The land area that drained into Hudson Bay and Strait, and into the waters off the arctic coast, had been called Rupert's Land, put under the commercial monopoly of the Hudson's Bay Company by British legislation in 1670, and in 1870 transferred from Britain to Canada, then just emerging from colonial status. Sovereignty over the regions farther north – Baffin Island and the rest of the huge, complex archipelago that stretched to within about 400 miles of the North Pole – whose geographical extent was known only in part even in 1903, was ostensibly British, based upon numerous voyages of discovery and exploration beginning with Frobisher more than three centuries before, and upon many individual declarations of possession, especially during the nineteenth century. In 1880 Britain transferred these territories to Canada as well, but the description of their limits was hopelessly vague, perhaps intentionally, and was therefore open to more than one interpretation. At first there was little concern in Canada, a nation of scarcely more than four million people struggling to cope with the political and economic

48 M.V.B. Millard, 'Journal kept by M.V.B. Millard on the Bark *Black Eagle* of New Bedford, Edwin W. White, Master, April 21, 1866 – Sept. 23, 1867,' Nicholson Whaling Collection, Providence Public Library, Providence, RI, 28 February 1867.

49 For more on the sovereignty issue see V.K. Johnson 'Canada's Title to the Arctic Islands,' *Canadian Historical Review* 14, 1 (1933): 24–41; W.F. King, *Report Upon the Title of Canada to the Islands North of the Mainland of Canada* (Ottawa: Government Printing Bureau 1905); G.W. Smith, 'The Transfer of Arctic Territories from Great Britain to Canada in 1880, and Some Related Matters, as Seen in Official Correspondence,' *Arctic* 14, 1 (1961): 53–73; G.W. Smith, *Territorial Sovereignty in the Canadian North: A Historical Outline of the Problem* (Ottawa: Northern Co-ordination and Research Centre, Department of Northern Affairs and National Resources 1963); Morris Zaslow, 'Gaining the Arctic Frontier,' *The Opening of the Canadian North 1870–1914*, 249–77.

problems of uniting its far-flung clusters of population between Atlantic and Pacific. The fortunes of the arctic islands were not of immediate importance to a people centred approximately half-way between the Pole and the equator. There were challenges and problems enough in the settled areas along the American border to occupy the attention of politician, industrialist, railway builder, farmer, and soldier.

Yet the nation had to take care that its generous geographical dimensions, so suggestive of resources for the future, would not be whittled away by the encroachment of foreign powers. The land, being vast, was vulnerable. In the late nineteenth century increasing awareness of the activities of foreign whalemen and explorers in the arctic archipelago, which Canadians understood to be under their jurisdiction by virtue of the British transfer of 1880, began to generate ripples of concern. In 1898 two foreign expeditions converged on the Smith Sound region between northwest Greenland and Ellesmere Island, one led by a Norwegian, Otto Sverdrup, the other by an American, Robert Peary. Both spent the next four years in the north. Peary's goal was the North Pole, and he attempted to reach it from Ellesmere Island on what he called the 'American route.' Several hundred musk-oxen, caribou, walrus and polar bear were killed to support his large expedition of sixty people and 200 dogs, without any consideration of the ecological impact. Sverdrup, also taking a heavy toll of musk-oxen, discovered several large islands west of Ellesmere Island, which were given his name and later claimed for Norway. The full implications of foreign involvement in the north now came clearly into focus. The sovereignty bestowed by the British order-in-council of 1880 was an insufficient deterrent to the ambitions of foreign individuals and nations. Jurisdiction over the arctic archipelago would never be secure until there was real and visible evidence of Canadian authority and administration in the north. This new point of view was supported by the unfavourable outcome of the Alaska boundary dispute. In 1903 an international tribunal brought down an interpretation of the vague 1825 Russo-British agreement concerning the position of the boundary between the Alaskan panhandle and Canada which effectively cut off the northern half of British Columbia from direct access to the sea.

The result was a hardening of the Canadian attitude towards the north. The alarmist view taken by some politicians was expressed by Senator Edwards to the prime minister.

In looking up the matter a short time ago, I was surprised to find the extent to which the Americans have been whaling in Hudson Bay and the many years they have been at it. Their aggressive and grasping nature is such that we need not be surprised if shortly they take the position that Hudson Bay is an open sea, and further, that they may lay claim to islands and territory in that North land, said to be rich in coal and a variety of minerals. It seems to me that we should lose no time in asserting our rights ...

If the Americans are permitted to skirt our Western possessions, for Heaven's sake do not allow them to skirt us all around. They are south of us for the entire width of our country; they block our natural and best possible outlet to the Atlantic; they skirt us for hundreds of miles on the Pacific and control the entrance to a vast portion of our territory, and the next move if we do not look sharply after our interest, will be to surround us on the North.[50]

Responding to the growing concern about Canadian rights in the north, the government dispatched an expedition to the eastern Arctic in 1903 to set up a police post in Hudson Bay, exact customs duties on any trade goods landed from whaling vessels, notify the native inhabitants that they were under Canadian authority, enforce law and order, and carry out scientific surveys (see appendix K). At the same time a police detachment was to be established in the western Arctic at Herschel Island, where the activities of San Francisco whalemen had caused considerable concern for the welfare of the Eskimos.

In Hudson Bay, the destination of Captain Comer and the *Era*, the geography of the coasts and inland regions was reasonably well known and Canadian sovereignty rested comparatively firmly upon two-and-a-half centuries of Hudson's Bay Company occupation of Rupert's Land and the transfer of these rights to Canada in 1870. But since 1860 there had been more than a hundred American whaling voyages into the Bay, during which the whalemen had reduced the Greenland whale to near extinction, brought the population of musk-oxen to a dangerously low level, and undermined the traditional independence of the Eskimos. The issue in this region was primarily one of regulating whaling activity, though the intention was also to strengthen Canadian sovereignty. One Canadian writer, J.B. Tyrrell, pointed out that 'it is just possible that, through long continued acquiescence, these foreigners [American whalemen] may be establishing rights whilst ours are being allowed to lapse.'[51]

The news of the forthcoming Canadian expedition on board the *Nep-*

---

50 Edwards to Laurier, 28 October 1903, Laurier Papers, Public Archives of Canada, Ottawa, MG 26, G.1(a), 228: 78415, 78416.
51 J.B. Tyrrell, quoted in P.T. McGrath, 'Whaling in Hudson Bay,' 198.

tune was not received with joy in New England whaling ports. American vessels had been whaling in Hudson Bay for four decades and during this time their freedom had been threatened only once: in 1866 the Hudson's Bay Company made a half-hearted and ineffective attempt to compete with the 'interlopers' upon their trading monopoly.[52] Since the transfer of the Company's jurisdiction to the Canadian government in 1870, the whalemen had operated without restriction and naturally resented regulatory measures. Although in fact the government intended merely, in the words of Prime Minister Laurier, to 'quietly assume jurisdiction in all directions,'[53] giving adequate warning of the new procedures and avoiding 'any harsh or hurried enforcement of the laws,'[54] some Americans chose to believe that the *Neptune*, a wooden sealing vessel chartered for the occasion from Newfoundland, constituted an 'armed expedition' whose purpose was to drive American whalers from Hudson Bay, using force if necessary. Force would be met with force, one writer warned and 'it will not be an easy matter for Canada to carry out the policy of expelling the alleged trespassers ... American whaling crews are not composed of the most tractable persons ... ' There might well be, he thought, 'clash between the two Anglo-Saxon peoples.'[55]

Pasted into the front of George Comer's journal are two newspaper clippings which reveal something of the curious public perception of, and reaction to, the *Neptune* expedition. The first, from a New Bedford paper, is undated but appears to have been published after the *Era*'s departure and pasted in later. Entitled 'Schooner *Era* in the trouble,' it records the comments of Charles T. Luce of the firm that owned the vessel. He is reported to have said, 'If there was no interference when there were 20 or 30 whalers in the bay, I do not understand why an expedition should be sent to drive out one little schooner. We have never done any trading in the bay.' His denial of trading activity was of course nonsense. The *Era* customarily departed for the north with an outfit of trade goods of which any Hudson's Bay Company factor would be envious (see appendix

---

52 The Company sent a whaleship north into Roes Welcome Sound, but either as a whaling voyage or as a move against the American interlopers the voyage must be considered a failure. W. Gillies Ross, 'Whaling in Hudson Bay. II. The Voyage of the *Ocean Nymph*, 1866–67,' *Beaver*, outfit 304, no 1 (1973), 40–7.

53 Laurier to Edwards, 29 October 1903, Laurier Papers, Public Archives of Canada, Ottawa, MG 26 G.1 (a) 228: 78417.

54 White (Comptroller RNWMP) to Moodie, 5 August 1903, RCMP Records, vol 293, file 236, part 2.

55 McGrath, 'Whaling in Hudson Bay,' 188.

c) and on her previous voyage (1900–2) had obtained from Eskimos the skins of 398 musk-oxen, 307 fox, 69 wolf, 51 polar bear, and 8 wolverine.[56] But Luce's remarks and other newspaper and journal articles suggest that New Englanders were not aware of the real intentions of the expedition. Indeed it is possible that its purpose may have been totally misunderstood in other areas outside Canada. The second clipping in Comer's journal is datelined Saint John's, Newfoundland (then a colony of Great Britain) on 29 September (presumably 1903). Its caption reads as follows: 'American whalers excluded. Canada orders New Bedford ships out of Hudson Bay. Claims waters are British and strangers must not enter. International complications likely to arise in connection with matter.' The article questioned the dominion's jurisdiction over the waters of Hudson Bay, stating that under international law the most that could be claimed as territorial waters was a three-mile coastal strip and, therefore, even if Canada's claim to have received the rights of the Hudson's Bay Company over Rupert's Land in 1870 from Britain were valid, the Bay itself – beyond the three-mile limit – was still a part of the high seas, and foreign vessels could not rightfully be excluded.

Although the Canadian government did wish to make Hudson Bay a closed sea (there was a move afoot to rename it the Canadian Sea), and succeeded in doing so a few years later, the *Neptune* expedition was never instructed to assert Canadian sovereignty beyond the three-mile limit, harass foreign vessels on the high seas, or expel American whalers from Hudson Bay. What was of practical importance to the whaling interests, however, was the nature of the restrictive measures to be implemented along the coasts, within waters indisputably Canadian, because by this time the whaleships were absolutely dependent upon contact with the land and utilization of its human and animal resources.

In retrospect, concern about the *Neptune* expedition appears to have been exaggerated, considering that only one American whaler, the *Era*, was to enter Hudson Bay in 1903, and that the restrictions imposed by the government really presented no serious impediment to her whaling operations. The truth of a situation is not always its most important ingredient, however. What was significant was that the whaling interests of New England – including the firm of Thomas Luce and Sons and Captain George Comer – did feel resentful. There is no hint in the official report of the *Neptune* expedition that the Canadian policy had been unhappily

56 The fur returns are totals obtained from daily entries in George Comer, Manuscript Journal on Board the *Era* 1900–2.

received, or that the captain and crew of the only American whaler in the Bay might have some cause to show irritation at the Canadian presence and the restrictions placed upon their activities.[57] Comer's journal, therefore, provides a most valuable commentary from a different point of view.

57 A.P. Low, *The Cruise of the Neptune, 1903–4.*

1

Voyage North
(29 June – 23 September 1903)

Monday June 29
Left home in East Haddam, Connecticut at 6 AM. I left home quite light-hearted considering what I knew lay before me and knowing that some familiar faces would be missing when I should return at the end of twenty-seven months, which is the time we are fitted for. My daughter brought me to the river in her team, she having married lately. Had to wait in New London [Conn.] about forty minutes where I met several friends who came to say goodbye. Had to wait in Providence [RI] fifty minutes and arrived in New Bedford [Mass.] a little after 2 PM where I was met by the old gentleman Mr Luce.[1]

The afternoon I spent in making the few purchases that I needed for the voyage and clearing the vessel at the custom house. Besides having a camera, a friend in New Bedford, Mr Charles Agard, fitted me out with another and twelve dozen plates (glass).[2] I have also a graphophone to take north with fifty blanks and about forty records.[3] These I believe will

---

1 The firm of Thomas Luce and Sons had purchased the *Era* specifically for Comer's first voyage as master in 1895.

2 Comer first used a camera among the Aivilingmiut and Qaernermiut of Hudson Bay in 1893–4, and within a few years was selling photographs of ethnographic interest to the American Museum of Natural History. There is a sizeable collection of his original glass plate negatives and lantern slides in Mystic Seaport, Mystic, Conn.

3 The graphophone, invented by C.A. Bell and C.S. Tainter about 1887, used a floating stylus to engrave grooves upon a rotating wax cylinder. The resultant cylindrical 'records' could be played back on the same machine. Thomas Edison's improved phonograph of 1888, rival of the graphophone, worked on the same principle; the distinction between them was largely one of name, and undoubtedly the names were sometimes used interchangeably. The American Museum of Natural History provided Comer with a recording machine (presumably an Edison Standard Phonograph for

be quite a surprise to the Eskimo. On my last voyage I made fifty plaster casts of the Eskimo faces.[4] This I expect to do again. Stayed at the home of Mr Thomas Luce where I have always been made at home. Raining during the afternoon.

Tuesday June 30
Weather improving, the wind coming to SW moderate. Have had a number of things given me while here and with what I have collected makes me out much more than I have ever had. After much delay in getting (and trying to get the men on board [see appendix B] we finally got started, leaving behind the man who had shipped as second mate and also the cook but got another man to go as cook, one who had been with me in the *Canton* (bark). The third mate took the place of the second mate and in this way we got away, a boatsteerer who has been with me now ten years to act as third mate. Took a snapshot picture of the men.

We are heavily loaded, much being on deck [see appendix C]. Got started at noon and the tow boat taking us well out by 3 PM we let go and then transferred the party who had come down the bay with us to see us off to the tow boat. The three voyages before this there have been a party of old men come down with us but this time they feel as though they were too old to come, including our Mr Thomas Luce. Fresh breeze

---

one is still in the possession of his grandson) and blank cylinders on this and subsequent voyages to obtain records of Eskimo songs and speech. About sixty-four of his phonecordings, made for the museum prior to 1909 are now in the Archives of Traditional Music, Folklore Institute, Indiana University, Bloomington. These sound recordings were the first ever made among the Eskimos of North America. Comer also took commercial recordings of popular songs north to entertain visiting Eskimos.

4 In the early days of physical anthropology plaster casts of the heads (and sometimes the feet and hands) of aboriginal peoples were systematically collected from many parts of the world to preserve a record of their distinctive physical characteristics, especially where traditional societies were faced by the uncertain prospect of rapid cultural change. A strong proponent of such collecting, particularly from arctic regions, was Franz Boas, anthropologist at the American Museum of Natural History in New York. In 1900 Boas persuaded Comer to obtain casts of Eskimo faces on his next voyage to Hudson Bay. A sculptor was sent to East Haddam to instruct him in the method of preparing the life masks, and three barrels of plaster were delivered to the *Era* before her departure. With his usual enthusiasm Comer set to work, and on his return two years later provided the museum with forty-nine casts of individuals from seven Eskimo groups encountered on the whaling grounds: the Aivilingmiut, Netsilingmiut, Iglulingmiut, Sinimiut, Padlimiut, Qaernermiut, and Sauniktumiut. On subsequent voyages to Hudson Bay he made additional casts. By 1917 he had deposited between two and three hundred at the museum.

from sw with irregular swell. Passed near No Man's Land. Divided up the men into watches and set the watch, the mate to take her out, as it is called, by having the first eight hours out. Hazy. Steering SSE. A number of the crew are seasick, including myself. Our second mate is the only one who is the worse for liquor, he being unable to remain on deck. One man, Mr Frank Borden, who came down with us, had been eight years with me. He cried because he was not able to go, being in poor health.

Wednesday July 1

Had moderate breeze from sw during the night. Partly cloudy and hazy. During the first part of the night passed over a fisherman's net which was set. In doing so it about destroyed the net. We passed near a fisherman and then hove to when he sent a boat to collect what was left of the net. Nearly all feel the effects of seasickness. Today there has been quite a little irregular swell. At 4 PM passed the large steamer *Patria*. She was steering E by N. The passengers gave us three cheers which we returned. Latitude at noon 40° 35'N.

Thursday July 2

The winds continue steady from the sw and weather remains the same – partly cloudy and hazy. Passed this morning in among a number of fishing smacks. They were fishing for swordfish. Two of our men are still unable to work, being very seasick. We find enough to do in stowing the deck load in better shape and getting the boats ready for use. We have four boats, three of them being new. The old one is a spare boat.

Friday July 3

Moderate weather. Winds west, varying in force fresh to light. Steering E by $1/2$ N. Distance run by log 132 miles since yesterday noon. One of our men claims to be a carpenter while another says he is a blacksmith. Such men are useful to have. Two men are quite seasick yet. Have seen a number of steamers today. Am quite well pleased with all our men and mates. We are not driving the vessel as fast as we might, till the men get more accustomed to the vessel. Only two of the men [seamen only] out of twelve have been at sea before.[5]

5 The ship's complement, after one mate failed to show up at departure, was nineteen all told (see appendix B). Seven were officers, boatsteerers, and steward, and twelve were seamen, or simply 'men,' of whom all but two were 'greenhands.' Experienced crew were scarce in the last decades of American whaling, and captains had to rely heavily on Eskimo labour once they reached the whaling grounds (see Elmo Paul

Saturday July 4
Very pleasant with fresh breezes from NW to north. At night wind increasing to a very strong breeze from north. Our sick men are improving
but not able to be around to work. The schooner leaks as bad as ever,
the water running in at the stem quite a stream. Seven PM put two reefs
in mainsail and took in fore topsail [see appendix A and glossary]. The
latitude 9 AM 42°32′N; the longitude 9 AM 62°00′W. Distance since yesterday noon 118 miles.

Sunday July 5
Fresh breezes during the night and this forenoon. Afternoon more moderate. Wind NNW, partly cloudy. As some water comes on deck we found
it necessary to make a new hatch cloth. Steering E by ¹/₂N. Distance run
since preceding noon 171 miles. Lat 43°35′N; long 58°32′W. Have one reef
in mainsail. No vessels seen the last two days. I have not got wholly over
being seasick yet. We are making a very fine run so far from home and
getting things stowed more where they belong. I am well pleased with the
appearance of the men. Four of the men [including officers] have been
with me before.

Monday July 6
Very pleasant. Moderate breeze from NW. Distance run since yesterday
noon 151 miles. Trying to get the vessel and boats into shape. The masts
are very black and all the spars are the same, not having been cleaned
since sometime last voyage. The vessel leaking so bad reminds me how
particular the owners are not to have her pumped out during the daytime
while loading and before going to sea, but have the vessel pumped out
during the night so as not to cause remarks. Our mate Mr Ellis does not
understand navigation but seems anxious to learn and for the safety of
the vessel I am doing all I can to teach him.

Tuesday July 7
Very pleasant most all day. During the morning had a few fog squalls.
The wind has gone around from north to east and is now SW. Distance
run by log 109 miles. We are now about forty miles from Cape Race [Nfld].
Lat 45°50′N; long 53°38′W. We still have one man who is unable to be
around to work. Each day finds us getting things into better shape. I think
that this is the finest weather I have had in all my trips along here, this
being my ninth time to go north.

Hohman, *The American Whaleman: A Study of Life and Labour in the Whaling
Industry* [New York: Longmans Green 1928]).

Wednesday July 8
Passed Cape Race at 8 AM and arrived at the entrance of St John's harbor at 6:30 PM but thick fog coming in at the time were unable to see or be seen. Wind moderate, north. The object in stopping here was to send letters home and hear what report there might be regarding ice to the north. We have now kept off, steering NE. There are two bergs in sight, one near Cape Spear, the other well offshore. Passed close to a small steamer which was fast to a finback whale. After killing it they towed it in to St John's harbor. I feel much disappointed in not being able to get our letters ashore.

Thursday July 9
After leaving last night and steering NE we found ourselves in near the land, the wind and current having set us in. When I was called at 2 AM to see the land so near and the swell heading us in, my mouth became perfectly dry so it became difficult for me to speak. We soon got the vessel heading offshore but this forenoon had both boats ahead towing. Put our letters on a fishing smack (*Maxwell*). He gave us some cod fish and I gave them tobacco and pork. Light air NE. Fog and rain during morning.

Friday July 10
The wind still remains NE moderate. At noon we were thirty miles south of St John's, the current running south at the rate of one mile an hour. Doing our best to get north. We have been carried back. Latitude at noon was 47°04′N. See quite a few fishing vessels. Quite a swell from the south.

Saturday July 11
Thick fog during the night and today. Wind fresh from NE to N. Heavy swell from SE. Last night we were very nearly run down by a bark (painted white). Today had to wear around to clear an iceberg. Six PM wind hauled to NW true, still fog. It is difficult to keep track of just where we are on account of such a strong current.

Sunday July 12
Thick fog during the night and today most of the time. Twice when the fog let up for a few minutes we got sights and by working them on the Sumner method we were able to find out where we are.[6] Lat 48°05′N; long

6 The Sumner technique in marine navigation was devised in 1837 by an American, Captain Thomas H. Sumner, to obtain a position line – a straight line on which the

50°23′W. Temp sea water 40°.[7] Nothing seen today. Started a new cask of water.

Monday July 13
Thick fog through the night. Had one reef in mainsail. Rough sea. Six AM temp water 40°. In rough seas the vessel leaks very much above the water line. One of our men who claims he is not able to go aloft on account of getting dizzy says he would not have come on the vessel but was told by the shipping agent that he would not have to go aloft. One other man tells the same story though he is able to go aloft. Wind moderate from NW true. Sea much smoother. Temp water 40°. Distance run since yesterday 114 miles. Nothing seen, partly from the fact that we can see but little ways on account of fog.

Tuesday July 14
Through the night we had fog till towards morning when it lit up for a short time, then shut in again till noon when we were to get our latitude, which was 52°01′N. Distance sailed since yesterday 125 miles. At night closed in again with fog. Nothing seen, either ice or vessels. Quite rough at times, wind NW fresh. Our seasick man is not able to be around yet. Temp sea water 42°. In stowing away things brought from home and friends it gives me much comfort to know I have so many good friends. Have started a fire tonight for the first time in cabin.

Wednesday July 15
Cloudy but no fog to speak of. This morning saw a steamer in the distance. Today we have had the boom and jib boom scraped, as we left home with all the vessel's spars in a very dirty black state, the owners not caring to spend money on such work and not having pride enough in the vessel's looks. Our latitude today was 53°04′N. Our longitude today was 49°39′W. This is a little farther east than we generally go in going north but it has been a week since we could get our longitude. Winds moderate from north, moderate swell. Last night I took my clothes off to sleep, not having had them off for the three nights previous. We have discontinued the use of sidelights as we are north of vessels going to Europe.

------

ship's position was located – using the sun's altitude and two or more supposed latitudes (Elbert S. Maloney, *Dutton's Navigation and Piloting* [Annapolis, Md: Naval Institute Press 1978] 571).

7 All temperatures given in the journal are expressed in Fahrenheit.

Thursday July 16
Cloudy with fog, squalls, light airs and calms. Moderate swell. Today
have been cleaning up the whaling spades and other gear and making line
tubs for the boats. Our seasick man is improving but not able to work.
Today has been our shortest run, only twenty-two miles. Lat 53°21′N;
long 50°00′W.

Friday July 17
Light airs and calms during the night. Today the wind is east, moderate,
with rain. Steering NNW true. Moderate swell. We have a very fine lot of
men and so far I am well pleased with the mate. Our second mate I hardly
think will amount to much as he has been a man who has drunk heavily.
Today we have commenced giving the men butter. I think I can allow
them ten pounds a month.

Saturday July 18
Moderate breeze from E to NE. Towards night nearly calm. Scraping the
masts and other work going on. Distance run last twenty-four hours 112
miles. Lat today 56°13′N; long 52°33′W. Pleasant towards night.

Sunday July 19
Cloudy during the day with fog squalls this afternoon, wind hauling from
WSW to north increasing to a strong breeze. Passed two icebergs this
forenoon. Lat 57°16′N; long 53°44′W. No work carried on on Sundays,
only what is necessary to work the vessel.

Monday July 20
Last night rather unpleasant but today weather improving, wind hauling
from NNW to WSW fresh. Steering north by compass and really going NW
as the compass is out four points in this latitude. We are now about 400
miles from Resolution Island, Hudson Strait. Got out a barrel of beef and
a cask of fresh water. Made a bread barge for the men. Temp sea water
43°. Lat 57°52′N; long 54°38′W. All goes well and pleasant.

Tuesday July 21
Very pleasant, wind moderate and hauling from WSW to NNW. Our latitude today was 59°52′N 55°13′W. This is much farther east than we generally have come but so much westerly winds have set us to the east while
trying to get north. We are now some 300 miles from Resolution. Today

we have finished scaping the masts as they had not been scraped since last voyage. This voyage we find that we are not as well fitted out or as much pains taken in stowing the vessel as last voyage.

Wednesday July 22
This forenoon fog with the wind hauling from WNW to NNW. This afternoon pleasant but at night set in fog again. Lat 59°45′N; long 57°38′W. Painted the top strake of all the boats black. Smooth sea and light winds. Temp sea water 41°. Several of the men have colds as though a distemper was among them. Two hundred thirty miles from Resolution.

Thursday July 23
Very pleasant last night and the morning but by 10 AM set in thick fog, wind hauling from north to east and is now SSE. A few bergs in sight this morning but the fog has hid everything from view. Steering NW by W (true), moderate swell from NE.[8] We are now 140 miles from Resolution Island, unable to get our position by the sun. Put on my woolen shirt today.

Friday July 24
During the night had the wind SE with fog. Kept going WNW (true) till midnight when we came to ice. Wore around and worked to the east till 7 AM when we ran to the north then WNW again till 5:30 PM when coming to more ice we wore around again. A few icebergs near. Fog at times quite thick. Wind fresh from SE. Resolution eighty-five miles off, WNW. Lat 60°50′N: long 61°00′W.

Saturday July 25
Last night we lay heading to the east on account of fog and the fear of meeting ice. This morning the land by dead reckoning was 100 miles off. The fog cleared up but the sky remained overcast. We ran in till 6 PM when the fog shut down again. We were now within twenty-five miles of Resolution. Have had a fresh [wind] from NE with a rough swell from south. Passed a few bergs. Took a snapshot of one with the camera.

---

8 It is assumed that the ship's course, the bearing of landmarks, and the direction of wind and waves are given according to the magnetic compass, unless specifically stated to be true, as in this case. The compass variation, or difference between true and magnetic north, was great in these regions and the erratic behaviour of the magnetic compass was a serious inconvenience to navigators.

Sunday July 26
The fog still continues with the wind SE to east during the day. The sun breaks out for a moment, when we get rather poor sights. Last night we headed to the eastward for fear of ice and today ran in towards Resolution Island. We raised it at 6 PM for a moment in the fog. Had to work to the south. Very little ice seen.

Monday July 27
The weather improved and the fog cleared away during the night. Today we are near the Grinnell Glacier which is on the north shore of the straits. Coiled down the whale lines and put up the crow's nest. Light air from east. No ice to speak of. A few bergs. Lat 61°33′N, long 66°55′W. I have been quite sick last night and today, having taken cold being up in so much fog as we have had and not undressing nights. It is such times as this that one longs for a woman's care.

Tuesday July 28
Sick abed with a cold and fever. Having been up nights as well as days and not having had my clothes off for several nights I think is partly the cause, but others have a bad cough and I have thought it might be a distemper among us. It is now the 31 of July that I am writing this, not having been able to be up till today. My people do all they can for me. The steward and a man (boatsteerer) we called Brass are affected, some much the same but no fever. We are near the Saddleback Island.

Wednesday July 29
Today have had a fever. No improvement in cold on the lungs. Seemed to be threatened with pneumonia as I could not breathe without hearing my lungs all the time. By painting my chest with Iodine and having two hot water bags and two hot bricks and then having them changed as they got cold, these made me sweat so that I felt better. Light airs and calms. We are still near the Saddleback Island. A few bergs but no field ice.

Thursday July 30
One month out from home today and I am sick abed but on the gain, but still cough considerably. The weather continues fine, light airs and calms, so we do not advance any to speak of and are still in sight of the Saddleback. Had tobacco dealt out to the men, which is the custom once a month. Each man is entitled to a pound.

Qaernermiut Eskimos on deck of the *Era*   Photograph by George Comer
Courtesy of the American Museum of Natural History

Aivilingmiut Eskimos on board the *Era*   Photograph by George Comer
Mystic Seaport Museum

Eskimos in *Era*'s 'house'   Photograph by George Comer
Mystic Seaport Museum

Eskimo group in *Era*'s cabin   The Whaling Museum, New Bedford, Mass

Friday July 31
This morning got up, had a change of clothes, and went on deck. Still have a heavy cough, but able to move around. Look over the number of packages which my friends sent to me and in opening one box which had in it several packages I came across a note from Mr Francis Parker, which I read with much interest, and seeing a paper of dates I began eating them and have now finished them. I have taken out for use today a little jar (it looks like pineapple) from Mrs Seymore. This I will have placed on the table at supper. A paper which I opened contained an illustrated letter from my girls, the misses Gillette, which brought a smile as each illustration filled in its part of praise.

At night we are close to Big Island or North Bluff. Have made but little progress the last few days. Now what wind we have is nearly ahead. In coming in to Hudson Strait have never seen so little ice. Latitude today was 62°15′N, long 70°30′W. We are now nearly due north of home. We had the jar of pineapple for supper. Pleasant weather.

Saturday August 1
We have light winds from directly ahead or WNW so that with the current out through the straits we advance but little. Possibly have made twenty miles during the last twenty-four hours, going from the north shore over to the south shore twice. We are now between Big Island and Spicer Harbor.[9] The weather has become overcast and raining lightly. Only a few bergs in sight but no small ice. This is an unusually clear season of ice. I slept very well last night, the first for some time. Still am not well yet of my cold. At 8 PM passed a log adrift. Lowered a boat and the mate went to see what it might be. Found it to be a trunk of a tree.[10] Could see where the limbs had been cut off and thought it might be fifty feet long.

Sunday August 2
Cloudy with moderate breeze from WNW which is directly ahead for us. Though we have no ice to contend with we have not got along well because

9 Spicer Island, on the north coast of Hudson Strait west of Big Island, supported a small whaling station between 1879 and the early 1890s.
10 Although the Arctic is characterized by treelessness, northward flowing rivers of the subarctic carry fallen trees to arctic waters where currents distribute them along many coasts. Driftwood was of great importance to Eskimo families, who used it for the frames of skin boats and tents, the runners of sleds, the shafts of harpoons and spears, and a number of other implements.

of light head winds. Have not seen the land today, Are making four-hour tacks. We are to the NE of Charles Island, a few bergs in sight. My cough is a little better.

Monday August 3
Fog squalls during the night. Today calm. We are near the south shore, about fifty miles east of Charles Island. This afternoon a breeze has come from the east. We pass a few pieces of ice but nothing to speak of. It seems very pleasant to have a fair wind as we have not had some coming into the straits. We have lost ground in the last twenty-four hours as the current set to the ocean. Filled up the scuttlebutt (water cask). This will last a week and we have enough for one more filling.

Tuesday August 4
Last night had a good breeze from SE but this morning the fog set in thick and, thinking I knew where the vessel was, [I] kept going, when at 11 AM we raised land ahead when the fog let up. We were able to just steer clear of it and work to the SSW. The island was Salisbury. I can feel that there is a God who watches over us. This has happened to us twice before this on this voyage, and many times in my past life. We are now 6 PM between Cape Wolstenholme and Nottingham Island. Wind WSW with quite a little swell. Have had some ice all day.

Wednesday August 5
Last night and today we have had light airs and calms with some fog and rain squalls. We have had a rough, irregular sea and have not gained any in our course. We can see the land both sides of us but are nearest to Nottingham Island, should judge about eight miles south of it. A very few pieces of ice to be seen. At 6 PM have a light breeze from SE and the sea has much improved. We are heading now for Coats Island. I noticed today that the men are wasting their bread so I have told them they should have less. The men are as a lot very good but slow to learn.

Thursday August 6
Last night had a good breeze from SE but this morning the fog set in thick again. We passed a few pieces of ice north of Mansel Island. Today at 1 PM made out Cape Pembroke, the weather having cleared. Have had calm and light winds from NE. We are now going along down the southeast side of Coats Island, Cape Pembroke in sight. Partly cloudy.

Friday August 7
We had a very good run down the SE side of Coats Island and turned to
the NW at 1 PM. Today we are now crossing from Coats Island to Southampton, hoping to be able to stop at the Scotch station which is in
63°09′N lat. From there we cross to Cape Fullerton. The wind is now very
strong from NE, though pleasant. If too much wind tonight will keep going
towards Cape Fullerton and not stop at the station. Have just put two
reefs in the mainsail. I have had a very bad cough lately and today I got
out a glass of the honey that Mrs Seymore sent to me, hoping it will help
my throat.

Saturday August 8
Last night had strong breeze from NE with thick weather part of the time.
When off Cape Kendall we touched bottom but did no damage. No land
in sight at the time but from aloft a little later land could be made out.
We came across and made the land half way from Whale Point to Cape
Fullerton, then ran down. Came to anchor at 1 PM. Found our natives
here have taken two whales, one eleven feet bone (later ten feet), and
one four to five feet. One of our natives had died while we were home.
This evening gave them a concert from the graphophone.

Sunday August 9
Very pleasant. Light winds, variable. Let the men take all their clothes
ashore to wash. Got two bear skins, one musk-ox skin and seven white
fox skins from the natives. Went around to all the native tents to see
them. In one tent where one of my natives had died, when I came in they
all began to cry as my coming in reminded them of their loss, knowing
that I would miss the man. There have been many deaths among the
natives in all the tribes. The natives of the Southampton Island, who
numbered fifty-eight people, have all died but one woman and four children. A number of the Kenepetu tribe and the Iglulic, while the Netchilic
tribe has lost a great many.[11] This is told me by my natives of the Iwilic
tribe.

11 The deaths among several Eskimo groups in northwestern Hudson Bay were evidently
    caused by the introduction of a virulent gastric or enteric disease to the Scottish
    whaling station on Southampton Island, and subsequently to Repulse Bay, by the
    Dundee steam whaler *Active* in July 1902. At both localities the presence of large
    numbers of Eskimos from various regions facilitated the spread of the disease (see W.
    Gillies Ross, 'Whaling and the Decline of Native Populations).

Monday August 10
Pleasant weather. Wind moderate from the NE to E. Commenced breaking out and landing provision, the natives helping us. As we have no one who can cooper the casks I have it to do.

Tuesday August 11
Fog part of the time with light winds from east. Have been landing provision and coal, also putting the copper and shoeing on the new boats.[12] Finished two of them. Two boats came this evening to the island but did not come to the vessel, it being late.

Wednesday August 12
Fog most of the time with the wind east. Two more boats came today from Whale Point.[13] Twenty musk-ox skins brought in today. Finished coppering the boats and putting on the shoeing. Have the covers most finished. We have landed nearly all our spare provision and now we are stowing empty casks to be filled later with fresh water or salt water.

Thursday August 13
Cloudy with rain squalls. Wind NE fresh. Finished landing provision. Got off three casks of water. We are lacking empty casks to stow in place of the full bread and flour casks. Took the raft apart.

Friday August 14
Cloudy, a heavy swell outside from SE. Got off two rafts of water, six casks. Took in trade today six musk-ox, two wolf, and two fox skins, and a little skin clothing and about twenty-five deer skins. Took a couple of snapshots of the natives while they were listening to the graphophone.

---

12 The whaleboats of ships wintering in the Arctic were normally employed in May floe whaling. Each night they had to be hauled out onto the edge of the fast ice and sometimes pulled onto ice floes when beset in close pack. To make hauling out easier their keels were shod with whale jaw-bone, and their bottoms sometimes sheathed with copper.

13 The boats referred to were whaleboats owned by Eskimos travelling to Fullerton Harbour to work for the *Era*. Comer had arranged on his previous voyage of 1900–2 that his 'ship's natives' should work on behalf of the *Era* during her absence, delivering their catch to the ship on her return in 1903, and continuing in his employment. Because of such arrangements Eskimo employment was reasonably continuous, despite the periodic departure of whalers.

Saturday August 15
Cloudy weather. Got off three casks of water this morning and three again this afternoon. Brought the boats off, having finished making the covers. We have got nearly through stowing casks and barrels. Took in trade nine musk-ox, three wolves, nine fox and two wolverine, also some deer meat and clothing.

Sunday August 16
Raining this forenoon. Got off a raft of water. This afternoon pleasant, wind NE. At night clearing, wind backing to SW. We have now about seventy-five natives around at meal times and am quite busy fitting the boats out for whaling.[14] We have six native boats in commission. I have quite high hopes of getting a few whales this fall. Two of our men had stole a box of tobacco and when I found it out I thought I should punish them but later after giving them a lecture I let them off.

Monday August 17
Light airs this morning but breezing up by 10 AM from SW. Got under way at 10 AM but before we got out set in thick fog, but we kept coming. Got up to Whale Point at 8 PM. Wind SSW, moderate. Bound for Repulse Bay. Have two native boat crews with us and four other boats are to go to Wager River.

Tuesday August 18
The weather improved during the night while we lay off and on at Whale Point.[15] This morning could see the house and the native tents on the point. We kept off up the Welcome with the wind south. We keep a lookout

14 When whaling masters hired ship's natives to assist in whaling, hunting, and other tasks, they undertook to feed their families daily, usually with biscuits, molasses, and coffee. American whalemen thus administered an early form of social welfare in the Canadian Arctic long before the Canadian government assumed any responsibility for Eskimo well-being.
15 There was no harbour at Whale Point but it was a popular rendezvous for whalers arriving in Hudson Bay and Eskimos seeking employment with the vessels. In addition, its position near the south end of Roes Welcome Sound made it a good lookout site and temporary depot, especially during the spring cruises by the *Era's* whaleboats. If the boats could not spare the time to transport baleen back to the vessel, still frozen in at a winter harbour, they could leave it at Whale Point in a wooden hut (twenty-six by sixteen feet) to be picked up later. Captain Comer also stored emergency food in the 'house' for ships' natives travelling along the coast.

now for whales. Have had a very good run up today. We are now quite a little above Wager River and near the place where the *Desdemona* was wrecked in September 1896.[16] Have had moderate breeze from south. We are now keeping the Southampton shore. Squally appearance.

Wednesday August 19

Had light winds the first part of the night from south then hauling from that to east and increasing. Came by Beach Point at 3 AM and came to anchor in Repulse Bay at 8 AM. Found the Scotch ship steamer here and the smack.[17] They have taken a small whale a few days ago, four foot eight inches bone. We found the bone of the whale here that our native (Harry) took last fall but believe that a part of it has been taken by some party. There is some heavy ice here. Wind NE all night.

Thursday August 20

Raining in the morning but becoming better later. Wind east and ice crowding up into the harbor. The Scotch people's natives got a whale today, about eight foot bone. We counted the slabs to the whale taken by our natives last fall and found only 540 when there should have been 640 slabs – evidently some one has taken the 100 slabs. In weight it was 1,500 pounds.

Friday August 21

Strong breeze from east which kept the ice moving through the harbor (heavy ice). The Scotch smack, not having any anchor but holding on to a box filled with stone made fast to a hawser, dragged close in to the

16 The bark *Desdemona*, another New Bedford whaler belonging to the firm of Thomas Luce, was wrecked on a reef about twelve miles north of Wager Bay on 2 September 1896. Captain Millard and his crew got off in three boats. They were picked up on the following day by the Hudson's Bay Company whaler *Perseverance* and were transferred to the New Bedford bark *A.R. Tucker* on 4 September. Two days later the *A.R. Tucker* put the men on board the *Era* for passage home. More than a ton of bone was saved from the wreck and some gear was removed by the *Perseverance*. When another salvage attempt was made in January there was no longer any sign of the wreck, pack ice having carried it off.

17 The *Active*, a 237-ton steam whaler from Dundee, had been making annual voyages into Hudson Bay since 1899 for the firm of Robert Kinnes and Sons. She normally exchanged supplies for produce at the mica-mining station at Lake Harbour, Hudson Strait, and the whaling and trapping station at Cape Low, Southampton Island, then cruised for whales in Hudson Bay until starting homeward in September. In 1903 the Southampton Island station was abandoned and the small ketch *Ernest William* (79 gross tons) was towed to Repulse Bay and later to Lyon Inlet to act as a floating base for whaling.

rocks. Helped haul her off and she is now made fast to the steamer. Our boats go off each day looking for whale while I stay on board getting things ready to go with the boats to Lyon Inlet. Both of the Scotch captains were on board this evening and I enjoyed their visit. There are sixteen boats go out each morning from all the vessels. Pulled out a tooth for one of the men and also one for a native. The smack's name is *Ernest William*.

Saturday August 22
This morning got the boat ready with provision for a ten days trip to Lyon Inlet. As I had more writing to do to send home by steamer *Active* I told Mr Ellis and Reynolds to go on ahead and wait for me.[18] After leaving the letters with the steamer's captain and giving to the smack's captain the use of two fluke chains and our kedge (as he has no anchor and has been using a large box of stone to anchor with); then leaving two men with the steward and cook, also Harry and Ben's boat crews to whale it from the schooner and to help take care of her, I started, but did not overtake the other two boats and hauled out at 9 PM inshore of a large island on the SE side of bay. Had quite a hard pull as the tide was down. Light winds and calms with a few showers.

Sunday August 23
Pushed off at 5 AM and having a light breeze from NE we soon came to where the other two boats were hauled out. Quite a pack of ice a little ways offshore. At ten stopped in a narrow place on account of strong head tide (going as we do, we keep inside the small islands). Started again at noon. Though quite a little ice we had no trouble with it. At 6 PM we got where we could look into Gore Bay four miles distance, but found our way blocked with heavy ice. The ice here and in Repulse Bay is known as Foxe Channel Ice and is generally covered with fine dust.[19] We hauled out at 6 PM to wait for the ice to slack up. Later wind strong from NE.

18 The *Active* was bound for Dundee but Scotland was often the most direct link between the eastern Canadian Arctic and the United States. Whalebone and discharged seamen were sometimes sent out by this circuitous route; equipment and supplies occasionally came in; and mail travelled both ways. Comer received fourteen letters when the *Active* returned from Scotland in the next summer (see journal entry 21 July 1904).
19 The sea ice of Foxe Basin has a characteristic light brown color. Explorers noticed it 350 years ago and it has been remarked upon frequently since. Wind deposition of dirt has often been suggested as an explanation of the discoloration but it appears more likely that fine marine sediments are stirred up from the extensive shallow areas of Foxe Basin by autumn storms, spread by tidal action, and incorporated into newly forming ice (N.J. Campbell and A.W. Collin, 'The Discolouration of Foxe Basin Ice,' *Journal of the Fisheries Research Board of Canada* 15, 6 [1958]: 1175–88).

Monday August 24
Pushed off at noon high tide and by keeping close in to the rocks we
managed to get about a half mile, then, waiting the tide half out and no
movement in the ice, we hauled out again. Light airs from south.

Tuesday August 25
The young ice formed last night on the rocks. The ice is packing in tighter
all the time so that it looks useless to try to get farther. In fact we can
not.

Have just found an old cooking soapstone kettle. It must have lain here
many years. It is slightly broken. There was quite a depth of moss over
some of the broken parts. I am walking over the same ground where we
were a year ago and it creates a feeling of sadness as I recall how much
I had hoped for and how little we got, then the death of one of our best
natives whose life might have been saved had we not tried to do so much.[20]
Since that time I have been home and stayed through the winter and am
now here again. A number of natives who were living then are now gone.

I think what a hard life these people have to live. They do not know
how much they suffer or are deprived of. I hope the time will come when
the civilized part of the world will do something to help them, such as
having a house of refuge once in every 200 miles where meat could be
cached when game was plentiful. Then there would be no periods of
starvation, and the loss of life among children would be much less.

We pushed off at noon and by keeping close in to the rocks we managed
to return and get back around the bend. Have given up trying to get to
Gore Bay and now we are working back to Repulse Bay. After coming
about ten miles we met a boat belonging to the Scotch smack natives.
They enquired if we had seen two other boats as there were three of them
when they left the smack but had got separated on their way. We kept
coming till 2 AM when we hauled out inside of the large island SE side of
Repulse Bay.

Wednesday August 26
By laying down the oars we slid the boats into the water at half tide. We
then came to the salmon stream and stayed for the night. On our way

20 In August of the previous year, while the *Era* remained at anchor in Repulse Bay, her
boats had cruised in the region of Lyon Inlet. One of the ship's natives, Paul, on
arriving at Lyon Inlet with additional provisions for the boat crews, reported that
several Eskimos at Repulse Bay had become ill and died. Shortly after Paul himself
developed a severe, bloody diarrhoea and died within a few days. This was evidently
the disease introduced by the whaler *Active* to Southampton Island and Repulse Bay.

here I stopped at the grave of the native who died about this time last year. His bones were not there and it was believed by the native who went with me that he had gone bodily to the spirit world. A box at the grave had in it a few things he had used in life, such as his pipe, tobacco, knife, and a revolver and cartridges that I had left for him last year, as he asked for them before he died. I left a jackknife there today.

Thursday August 27
Stopped at the salmon stream last night and left there at 4 AM. Worked along slowly towards the vessel and arrived on board at 7 PM. The steamer has taken a medium sized whale since we have been gone. She is still here. Light airs and calmer.

Friday August 28
Pleasant. Light airs and calms. Have been off with the boats over towards the Blue Land.

Saturday August 29
Got under way at Repulse Bay this 11 AM. Wind SW moderate but increasing, a smoky southwester. Quite a little ice to come through. We are perhaps ten miles south of Beach Point. This morning the Scotch captains let me have the keys to the house on Southampton, in case I should care for anything which he had left there.[21] I gave him a few boards which he had.

Sunday August 30
Wind fresh from the SW. We worked down to Wager River where Mr Cleveland came off to us, some natives bringing five musk-ox skins, fourteen fox skins, [one] wolf skin – a poor lot. I gave him some tobacco, bread, butter and other things.[22]

21 The Scottish firm of Robert Kinnes and Sons of Dundee had established a station for whaling and trapping near the southernmost extremity of Southampton Island in 1899. It was staffed by three Scots and more than 100 Eskimos. When the station closed in 1903 the men left the station house and a lookout tower standing.
22 George Washington Cleveland was landed at Wager Bay by the American whaler *Francis Allyn* in the summer of 1900 to establish a small whaling and trading station for Thomas Luce and Sons. Independent in spirit and somewhat erratic in behaviour, he appears to have attached little importance to his relationship with the company. In August 1904 he signed on to the small Scottish ketch *Ernest William* at Lyon Inlet, leaving almost 400 rotting skins at the abandoned station at Wager Bay for Comer to pick up (see journal entry 14 September 1904). In 1921 Cleveland was still in the region, serving as manager for the Hudson's Bay Company post at Repulse Bay,

Monday August 31
Strong breezes from SW. We are about fifteen miles south of Wager River. Some of the natives we have on board are quite seasick. Have one reef in mainsail and fore topsail furled. Some fog and rain.

Tuesday September 1
Weather improving, with less wind. At night calm. The natives went ashore near Mount Zion, which is north of Yellow Bluff. Wind still NW.

Wednesday September 2
Light airs during night hauling to NE. Kept off at daylight and came to Whale Point when by that time we had a moderate gale with a little rain. We kept coming towards Cape Fullerton but missed the entrance. Sleet and rain, wind hauling from NE to NW. Laying to in a very rough sea. Boats in upper cranes. Weather improving at dark.

Thursday September 3
The weather improved during the night, wind hauling to west and SW. Came up by Cape Fullerton to Whale Point when I had the boats go ashore and bring off the musk-ox skins which were in the house there, some fifty of them, also four bear skins. Sky becoming overcast from SE. Light winds, smooth sea.

Friday September 4
Pleasant with a few rain squalls, wind south fresh. Spoke the Scotch steamer from Repulse Bay. She was unable to stop at Wager River, as her captain had intended, to take Mr Cleveland and his goods home with him. Finished writing letters and put them on the steamer.

Saturday September 5
Light moderate weather. Light air SW. We are off Whale Point.
A few light rain squalls during the night and forenoon. The native boats came off to us, Harry and Ben going ashore at 8 PM. Fine weather. No whales to be seen. The natives brought off $1^1/_2$ saddles of deer meat.

Sunday September 6
Light winds during the night and this morning, increasing to a gale from

---

'an old whaler who had been stranded on the coast here over a generation before, and made himself so comfortable among the Eskimos that he had never been able to tear himself away' (Knud Rasmussen, *Across Arctic America. Narrative of the Fifth Thule Expedition* [New York: G.P. Putnam's Sons 1927], 9).

the sw. Made an effort to get in to Cape Fullerton but could not reach there before the sea got too rough. Are heading off s by E. Heavy irregular swell.

Monday September 7
The weather improved during the night, wind hauling to west and moderating though the swell has continued all day from north. Rove off new flying jib sheets. At night are half way between Cape Fullerton and Whale Point.

Tuesday September 8
Moderate through the night, wind hauling to NE and the weather looks threatening. Came to Cape Fullerton and anchored at 4:30 PM. Harry's and Ben's boats are here. Sky overcast and strong breeze from ENE. Quite a swell from SE. Let go second anchor at 7:30 PM.

Wednesday September 9
Very heavy gale from NE with snow this forenoon. This afternoon stopped snowing and clouds more broken. Have out forty-five fathoms on the large anchor and thirty on the small anchor.

Thursday September 10
Still blowing heavy from NE hauling to north. More moderate towards night. The snow has formed in drifts on the land and all day keeps dropping from aloft down on deck. Repaired the flying jib and bent it again. Opened a barrel of beef.

Friday September 11
Still blowing strong from NW, a few snow squalls. Landed ashore some deer skins we had taken from Whale Point and brought off two empty casks to fill with musk-ox skins, as we had taken fifty-six from Whale Point. Landed a pair of whale jaws which we brought from Repulse Bay, these to be used in shoeing sled and boats. Repaired the jib topsail.

Saturday September 12
Blowing strong from the NW. The natives were unable to come off. Shifted water from forward casks to those under main hatch. Snow and rain squalls this afternoon. Mr Ellis hurt his back today, sometimes known as dropping a stitch.

Sunday September 13
Blowing strong from NE with light rain. Our four boats came from Whale Point this forenoon. They have been near Wager River and Southampton Island but did not see any whales. Mr Ellis's back is quite lame. Put on Iodine and later a Paris plaster. At night weather improving. Hove up the second anchor.

Monday September 14
Sky overcast with fresh breeze from NE to NW variable.
The natives shot fifteen ducks this afternoon. Had the graphophone out this evening. As the water freezes snow on deck we took the faucet out of the water cask.

Tuesday September 15
At 5 AM got under way and came out with a fresh breeze from NW. Went down in sight of Depot Island then turned and came up in sight of Whale Point by sunset. The weather looks northeasterly and now we are heading to the SW. Cloudy this forenoon, afternoon clearing. The high hopes that I had when we first arrived have dropped very much, as we have not seen a whale since we have been here. The weather is getting quite cold.

Wednesday September 16
Wind WSW during the night and this forenoon. Snow squalls all day. Wind increasing and hauling to SW. We worked down to Cape Fullerton and came in to anchor at 6 PM.[23] Blowing strong from SW. Today we put a reef in the fore topsail. It is the first time that it has ever been reefed since I have been in this vessel, and while a whole topsail would have been too much sail, after it was reefed it helped very much in tacking in such a rough sea.

Thursday September 17
Rain and fog with fresh breeze at first, then moderating. Wind SW. At night thick fog and calm. Came in to the inner harbor this morning, where we wintered three years ago. Got two rabbits and one duck.

23 Fullerton Harbour is located in Bernheimer Bay just west of Cape Fullerton among
   several islands, including Major Island, Beacon Island, and Store Island. The protection afforded by the island group and the easy access over landfast ice to the mainland were important advantages for wintering whalers. Comer had wintered here
   in 1895–6, 1897–8, 1898–9, and 1900–1; in his journals he refers to the place as 'Era
   Harbour.'

Friday September 18
Thick fog with moderate breeze from east. The boats got nine ducks today.
Have been making a bucket.

Saturday September 19
Strong breeze from SE to ESE. Have shifted the chain lockers from near
the mainmast to the foremast.

Sunday September 20
The weather coming on very bad, wind strong from east. Rain first then
snowing thick. Some swell gets in here, but very slight.

Monday September 21
Blowing heavy from ENE, some snow. Laying by both anchors. Tonight
weather slightly improved though still blowing heavy but not steady. Built
the chain lockers near foremast.

Tuesday September 22
Still blowing heavy but weather improving. Put our boats ashore for the
winter and are using two old ones.[24] Four rabbits and two ducks and a
pail of mussels were got today. Wind NE.

Wednesday September 23
Weather more moderate. Wind still from NE. Sky overcast most of the
time. Unbent the mainsail and have been putting an extra pump down
through the galley. This afternoon the steamer *Neptune* employed by the
Canadian Government came in here acting as a revenue cutter [see appendix K]. She is going to winter here with us and it looks as though she
has brought seven men known as Mounted Police to watch that we did
not land anything ashore that they claim we should pay duty.[25] I went out

24 Whaleboats and other gear were stored ashore during the winter in order to make
more room aboard the vessel and to provide security against possible loss of the ship
by fire or other causes.
25 Comer says there were seven policemen on board the *Neptune* but according to Low's
account there were only six: Major Moodie, Staff Sergeant Dee and constables Tremaine, Jarvis, Connelly and Donaldson (A.P. Low, *The Cruise of the Neptune, 1903–4*
xvii). The official report to the Comptroller of the Royal North-West Mounted Police
does not list the personnel of the detachment but does mention Dee, Tremaine, and
Conway, whose name appears to have been misinterpreted by Low as Connelly
(J.D. Moodie, 'Report of Superintendent J.D. Moodie on Service in Hudson Bay ...
1903–4,' 5).

to her as soon as we knew she was coming in, though by that time she was more than halfway up the entrance, and piloted her the rest of the way to an anchor near us.

Took dinner on board 2 PM. She has a crew of forty-three men all told. Major Moodie has charge of the police, Captain Bartlett of the steamer, and a man by the name of Low has charge of the expedition. He has gone up into Chesterfield Inlet for natives, with a launch.

2

Taking up Winter Quarters
(24 September – 31 December 1903)

Thursday September 24
Moderate breezes from NE, light at times. The steamer's party landing lumber to erect a house on an island where we get water. Our people got eight ducks and three rabbits today.

Friday September 25
Moderate weather, wind from NE to NNW. Our people got seven ducks and two rabbits today. We brought off our lumber to build our house over the after part of the vessel.[1] I have felt quite disturbed about the steamer coming here to establish a station, but conclude we can do no different than to remain here and continue on as we should have done if she had not come. She will of course collect many natives around here and this will take much clothing and some trade which would have come to us.[2]

---

1 Building a 'house' on the deck of the vessel was a standard wintering technique in arctic regions. It provided extra living space in ships designed primarily for cargo capacity and seaworthiness rather than comfort, and it was often used for plays, concerts, and dances, as well as for such practical tasks such as repairing boats in cold weather. On the *Era*, a schooner, the house was built aft of the mainmast, using the main boom as a ridge pole. It was constructed of wood planks and canvas taken north for this purpose (see appendix C).
2 Comer's concern that the government steamer would compete with the *Era* for fresh meat, winter clothing skins, and commercial furs was entirely justified. The crew of the *Neptune* (almost three times as large as that of the *Era*) were fully equipped with skin winter clothing, provided with fresh caribou meat through the winter, and even 'encouraged to hunt or attend lines of fox-traps for exercise' (A.P. Low, *The Cruise of the Neptune, 1903–4*, 28). Men of the police detachment operated traplines as well. Furthermore, Superintendent Moodie sent sleds to Chesterfield Inlet and Baker Lake more than once to trade with Eskimos of those regions (see journal entries

It will also make game less in proportion, as there will be more natives to hunt.Our natives are not having much success so far in deer hunting, at least those we hear from. Took up our second anchor.

Saturday September 26
Moderate weather. Commenced building the house over the after part of the vessel. The steamer's party have got the framework of their house up. So far none of them have called here on the *Era*. Wind light from NE. Our people got seventeen rabbits and one duck today.

Sunday September 27
Moderate storm from NE. Our people got twenty rabbits and two ducks today. The mate of the steamer came over today and the ship's carpenter, and had dinner. Some Kenepetu natives came today, brought one bear skin, two wolf, and one wolverine, and also ten deer skins.

Monday September 28
Moderate weather, though during the night we had quite a fall of snow. Got along quite well with the house. This evening the mate of the steamer, Mr Hearn, the doctor, and Mr Caldwell the prospector, and a Mr Borden came over and spent the evening. We enjoyed the visit very much. Wind NE, moderate.

Tuesday September 29
Wind hauling to ESE and coming on bad weather, snowing quite thick. Our people got ten ducks and four seal today. Covered the house with old canvas. One wolf skin and a number of deer skins brought in today. We have now collected about seventy-five deer skins for winter clothing and spring use. If our natives are not successful we will let them have some of them. These skins we have got from outside natives who have not been in boats looking for whales.[3]

Wednesday September 30
Fresh breeze from SW with snow squalls through the night and more

---

10 December 1904, 6 January 1905). Furs obtained by the *Neptune* and *Arctic* expeditions and the police, including almost 400 white fox, were sold at auction in 1906, for $1,913 (RCMP Records, vol 320, file 545).

3 Outside natives were those not employed regularly by the ships, who often visited from distant localities, bringing a few articles to trade, and sometimes remained at the whaling harbours for extended periods. They were a useful labour reserve and supplementary source of meat and skins.

moderate today. Working on our house over the vessel. The man who has charge of the Mounted Police, Major Moodie, called for a few minutes today. I was quite pleased with his call.

Thursday October 1 Steamer *Neptune*

Snow squalls with the wind NW. Word came this morning from Chesterfield Inlet to the steamer that the steam launch which the commander was cruising in was wrecked at a place known as Dangerous Point. The steamer at once got up steam and her captain, Bartlett, sent word to me to come over and see him. While doing so he wished I would accompany them, as I might possibly be of service. Telling him I would, I returned and got ready. Taking three of our natives with me and Brass my boatsteerer and one of our whale boats, we got started at 11 AM high tide. At 6 PM came to anchor south of Depot Island in ten fathoms near the shore, the wind being strong from NW with snow squalls.

Friday October 2

At daylight we got started and headed for Chesterfield Inlet, but in keeping near the land we struck twice on reefs, the reefs making well offshore at the north side of the inlet, say eight or ten miles. Then passing between Fairway and Promise Island we had fourteen fathoms, then into deep water between these islands and the south shore. At night came to anchor at a point above Merles Harbour, ten miles below Dangerous Point. We were seen by the commander (Low). Sent a boat but it got confused and came back without the party.

Saturday October 3 *Neptune*

This morning at 6 AM started from the steamer with our boat and natives, and came up to where the launch was wrecked. Found the party of four all well. We went out at once to the launch, it being low tide. Got ropes ready fast to her so that when the other boats came we could raise her. By that time they began to arrive and by lashing our masts and small spars across the boats we floated and warped the launch to a higher place at high tide, hoping to be able to patch. The tide being high at noon we could do no more till low tide. The weather set in stormy at noon, increasing to a gale from the northeast. The commander and Captain Bartlett with about ten men started to go to the steamer but fearing they would pass her in the storm they stopped and remained on the land about halfway between us and the steamer and passed a very uncomfortable night, reaching the steamer next day at 11 AM.

Commander Low requested me to remain in charge of the party at his camp. There being fourteen of us and too many for the tent, we turned over one of the boats and putting the sails up to help keep out the snow, seven of us slept under that. Storming too hard to do anything to the launch. We had plenty to eat and were comfortable.

Sunday October 4 Living under a boat at Chesterfield Inlet
We are having a very heavy storm from NE. Can see but little ways through the driving snow. Impossible to work out, and so we eat and sleep. Under the boat with me are the four natives Gilbert, Sam, Ben, and Scotty, then the sergeant of the Mounted Police and the steamer's carpenter, while the other seven men (four Mounted Police, a fireman, and an engineer and a foremast hand) make out the other seven men who sleep in the commander's tent. We have plenty of hardtack and corn beef. We are in a small cove on the north shore mainland.

Monday October 5
The wind hauled from NE to south then SW during the night. Still blowing and snowing heavy. We went off to the launch at 7 AM, found that we could not patch her up so as to float her. Her starboard side is much broken and now some stones have come up through her port side, but could not tell how much she is damaged on that side. Would return to the steamer but are unable to move on account of the storm and wish to take all the camp things with us.

Tuesday October 6 Steamer *Neptune*
The weather improving, wind west still blowing. We broke up the camp and brought nearly everything to the ship, and made our report of the condition of the launch. Found all well and we received dry clothing also, the natives also.

Wednesday October 7
After getting the second anchor clear, got under way and sounded our way up to where we had been camped and came to anchor, it being too late for the low tide. Got the empty tierces and bags of coal from the camp, and got ready for tonight's tide.

Thursday October 8
We worked and raised the launch on the midnight tide last night and got her to the steamer at 7:30 AM. We floated her by putting empty tierces in her and fastening them down and then a boat on each side with spars across her. Had got her chains shackled on for hauling. At low tide took

her on board by steam derrick and at 2 PM got under way and came down nearly to Wag Island and anchored for the night, wind NE blowing strong at night.

Friday October 9
Blowing strong from NW, the wind having backed from NE. Laying with one anchor and seventy fathoms of chain out in twenty fathoms of water. The wind drawing out makes quite a little swell. At night a little improvement in the weather.

Saturday October 10
Weather improved a little but still snow squalls with strong winds from NW. Started before daylight and came out between Promise Island and Fairway Island. Kept well out, then headed up and made the land north of Depot Island. Got to Cape Fullerton at dark and came up and anchored in the outside harbour at 6 PM. The natives and Brass took our boat and went to the schooner to remain here as we intend to go inside in the morning and I am acting as pilot.

Sunday October 11
Pleasant weather, wind fresh to moderate NW to N. Came inside where the schooner is anchored with the steamer. I stayed to dinner and have been invited to come every Sunday during the winter. This evening attended divine service, which was conducted by Major Moodie, the service being that of the Church of England.

Monday October 12
Wind ENE to SE fresh. Young ice making. Cut our winter supply, the steamer people doing the same while Major Moodie and his party are at work on the house which they are building ashore near our pond.[4]

Tuesday October 13
Partly cloudy, light winds ENE. Good weather to work outdoors. Our carpenter has been working on the house for Major Moodie, and I have been helping. The steamer's men have been cutting ice on the pond. Young ice is making in the harbor but is kept moving by the wind and tides.

---

4 Pond ice was the customary source of winter drinking water for arctic whaling ships. Ice blocks were cut in the autumn, sledded to the harbour, stacked near the vessels, and melted down as needed (see Comer's remarks on water in appendix J).

Wednesday October 14
Wind SE, stormy most of the time. Let go the second anchor at 3 AM, some drift ice floating by.

Thursday October 15
Blowing strong from the NW with snow. Hove up and let go again the second anchor. The natives are unable to get ashore.

Friday October 16
Pleasant though partly cloudy. Weather getting colder – this morning five above zero, at night two above. The harbor is covered with young ice. Got one duck. Wind moderate NW.

Saturday October 17
Partly cloudy with a little snow. Wind this morning SE, hauling to south, then west and NW and north, moderate. Some more of our natives came today and came off on the ice to the vessel, though it is quite thin. Harry, Billy and Tom Nolyer, with their families, were the people who came today. A party of the crew of the steamer came over this evening, including two of the policemen. Built the porch or outside doors today. Temp 2° above 7 AM, 3° below 6 PM.

Sunday October 18
Pleasant though partly cloudy. Wind moderate NNW. Took dinner on board *Neptune* and Commander Low returned with me and took supper with us. This evening attended services on board the steamer. The ice is now strong enough to go on with safety.

Monday October 19 Cape Fullerton
Today we took up our anchor and by sawing we placed the vessel in position for the winter heading due north, the ice being now four to five inches thick [see appendix F].[5] Wind NW, blowing quite heavy.

5 At freeze-up Hudson Bay whalers were usually positioned heading north in order to receive as much sunlight as possible through the windows of the after cabin (where the captain lived), to give each side of the vessel an equal duration of sun, and to enable the snow banking around the ship to melt equally fast on each side of the hull in spring. In the western Canadian Arctic, on the other hand, the practice was to place the vessel parallel to the prevailing storm winds to avoid snow drifting in heavily along one leeward side (J.A. Cook, *Pursuing the Whale, a Quarter Century of Whaling in the Arctic* [Boston: Houghton, Mifflin 1926], 57). After a ship was positioned, her

Tuesday October 20
Blowing heavy with some snow from WSW to west. Afternoon weather clearing but still blowing, wind west. Got out bread, 475 lbs. Got out flour. Temp 7 AM 8° above. The vessel does not leak as much now we are frozen in.

Wednesday October 21
Strong breeze from west, moderating towards night. We helped the steamer's people cut the ice and haul their vessel into position for the winter. Stayed to dinner. They intend to print a small paper occasionally and ask to have us contribute to its pages. Went over this evening and handed in my notes on the Eskimos.

Thursday October 22
Moderate breeze from NW to WSW. Today I have had a slight attack of pain in my bowels. The Captain Bartlett of the steamer came over to call and when he returned he sent the doctor over. Later Commander Low came over. Our carpenter is at work on the house for Major Moodie. More of our natives came today – Melichi and Stonewall.

Friday October 23
Forenoon pleasant. Afternoon wind getting to SE and stormy. Took our last boat ashore for the winter. Am feeling better today. Last night our natives had one of their times or anticoot meetings and claim to have killed an evil spirit which has been causing sickness around here.[6] On

---

anchor was raised and taken on board; the firm harbour ice would hold the vessel until spring break-up (see Comer's remarks on freezing a vessel in in appendix J).

6 Eskimo shamans – sometimes called angakoks (variously spelled), medicine men, conjurors, or witch doctors – held mystical meetings from time to time, during which they communicated with the spirit world in order to appease spirits for broken taboos or to obtain help in times of illness, hardship, lack of success in hunting, and so on. These séances, festivals (as Boas called them), or anticoots (the term used by Comer) were a vital part of the shamanistic rituals. Comer respected the native beliefs and ceremonies; the anticoots were sometimes intended to convey benefits to the whalemen as well as to the Eskimos, as his earlier experience illustrates:

Last night the natives held a meeting and one of them who pretends to have power to visit the spirit land went into a trance to call upon the spirits who have taken charge of the whales and learn why we had not had better success. This morning the natives told me that the anticok was told by the deities who look after the whales that we had worked on the musk-ox skins at a season when we had not ought to, or perhaps we worked on them on the vessel and that being on the water

board the steamer they have printed a small paper (3 sheets of typewriting paper), its first edition coming out today. It is called the *Northern* ————.[7] Sixteen salmon brought in.

Saturday October 24
Stormy. Wind SE to NE, then NNW. Afternoon better weather, though not pleasant. Had an entertainment on board here this evening. The people of the steamer *Neptune* were over. We are about 500 feet apart. Fifty-two salmon brought in today.

Sunday October 25
Fresh breeze from NW. Pleasant weather. Temperature 18° above this 8 AM, growing a little colder this afternoon. Three more of our natives came this evening, have done quite well in deer hunting. Took dinner on the steamer and attended divine service the evening. Commenced on two meals a day. Three more of our natives came.

Monday October 26
Partly cloudy. Wind moderate NNE, some snow. Brought off some cakes of ice which we intend to put over the windows later. Three seal got today. Morning temp 12° above. Mr Ellis went off fishing, got six salmon.

Tuesday October 27
Fresh breeze SSE to south. Weather warm, thermometer 28° above. Four seal got today. Overcast. Got off two loads of ice for water and a load of wood.

Wednesday October 28
Mild weather with rain this morning, the wind SW then west. Afternoon wind NW and becoming colder. Morning temp 28° above, afternoon 18°

---

was wrong. Also we had an owl skin which we had saved and we had done wrong again in picking the feathers off the ducks. We had done wrong, we should have skinned them. I was told to throw up my hands several times as though throwing things away. After I had done that (which I did) they told me I was alright now and would have good success in future. (Manuscript Journal of George Comer on Board the *Era* 1897–9, G.W. Blunt White Library, Mystic Seaport, Mystic, Conn 14 September 1897).

7 The name finally adopted for the newspaper was the *Neptune Satellite* (see journal entry 16 November 1903). No surviving copies are known to the editor but conceivably some exist in family collections.

above. Took dinner on the steamer with the scientific people, who eat by themselves.[8] Nine wolf skins and three deer saddles brought in.

Thursday October 29
Pleasant and much colder. Thermometer 2° above, at night 2° below. Two rabbits shot today and one partridge. Also four salmon taken. While taking pictures today my camera got out of order and spoilt the whole twelve plates – all the pictures I had taken. Wind NW moderate. At night calm.

Friday October 30
Sky overcast, wind NE moderate. My eye has something in it which makes it quite painful. The Doctor Borden of the *Neptune* had his foot badly hurt while drawing off ice. Three rabbits got today. Opened a can of Mrs Seymore's honey.

Saturday October 31
Moderate breeze SSE with a slight fall of snow towards night. My eye is a little better but am not able to be out in the light. Five salmon and one rabbit today. The paper issued by the people of the *Neptune* came out again today. All of the men forward now have deer skin coats.[9] Have charged each one a dollar apiece for them.

Sunday November 1
Wind south. Overcast with a little snow. Two sleds (Kenepetu) came to the steamer – Molasses and Anticoak. This man Anticoak is one who I let have a rifle and loading tools with enough ammunition to last him all winter, he to pay me fifteen musk-ox skins. Now he seems to have forgotten me and is trying to work for the steamer. My eye is a little better but still

---

8 The scientific party on board the *Neptune* numbered six: A.P. Low (geologist) L.E. Borden (surgeon and botanist) G.B. Faribault (assistant surgeon) A. Halkett (naturalist) C.F. King (topographer and meteorologist) and G.F. Caldwell (photographer). Low, *The Cruise of the Neptune, 1903–4*, xvii.

9 Comer always hired Eskimo women to make up caribou skin clothing for the crew. Without it they would be ill-equipped to work, travel, hunt, or play outside in winter. His example was followed by Commander Low. Clothing for the two crews – fifty-seven men altogether – probably required over 300 caribou. The women made up boots as well, probably of bearded seal, ringed seal, and caribou skin. In the back of Comer's original journal is a list of eleven Eskimo women who made fourteen pairs of boots for the *Era*'s crew in the spring of 1905.

the light hurts it. Have not been able to go to the steamer to dinner as usual or attend services in the evening.

Monday November 2
Mild weather though the wind has been west. Snowing at night. Today we have begun making sleds for the natives. The mate of the steamer and Mr Caldwell the prospector have been here and spent the evening.

Tuesday November 3
Wind west, varying in force. A little snow falling, though partly pleasant. We are at work on three sleds for the natives. Three salmon brought in today. My eye still troubles me.

Wednesday November 4
Quite pleasant weather, wind moderate NW, partly cloudy. Two rabbits got today. I have been over twice to see Doctor Borden on the steamer. He has picked the lids of my eye and put on a bandage, first putting on tea leaves. They have another doctor who is French but is either insane or has softening of the brain.[10] They are preparing to confine him, as he is not safe to be at large. Finished the three sleds. As we have not got whale jaws to spare to shoe them the natives will put on moss.[11]

Thursday November 5
Wind west this morning but now south and increasing in bad weather. My eye is a little better, the Doctor Borden doing what he can for it. I go to the steamer to have him attend to it. He is now able to walk. The other doctor has to be kept under guard of one of the police, of whom there are six including the major, who is appointed governor of all this country from Fort Churchill north.[12]

10 G.B. Faribault, MD, was the *Neptune's* assistant surgeon. Soon after leaving Halifax he began to show signs of what Low called 'mild insanity' and in Hudson Bay his condition became much worse. He would sweep the ship's decks all night, answer imaginary telephone calls, and behave in other deranged ways. When he set a fire inside his cabin in order to 'sterilize' it, Dr Borden certified him insane and from November 1903 until his death on 27 April 1904 he had to be confined in a makeshift cell and guarded, dressed, washed, and exercised by police constables. The cause of his insanity was said to be syphilis acquired years before as a post-graduate student in Paris (Lorris Elijah Borden, 'Memoirs of a Pioneer Doctor,' 72, 74).
11 The usual shoeing for sled runners was whale jaw-bones, baleen ('bone'), or ivory. A coating of ice on top of such base materials made a fast surface. For unshod wooden runners wet moss was plastered on and then iced (Franz Boas, *The Central Eskimo* [Lincoln, Nebr: University of Nebraska Press 1964], 122, 126).
12 Major Moodie was appointed superintendent commanding 'M' Division of the Royal

Friday November 6
Cloudy with a little snow and fresh wind from SW. Got off two loads of ice and got out ten boxes of bread.

Saturday November 7
Quite mild weather, wind moderate from NW. Took a number of pictures of the schooner, the steamer, and natives' snow houses, my eye is much improved. Had the graphophone out the evening and got a couple of records of native songs. One of them was of the Kiackennuckmiut. The other was Iwilic.

Sunday November 8
Very pleasant, light air from north. Took dinner on board the steamer. They have now commenced on two meals a day – breakfast at ten and dinner at 4 PM. Attended service in the evening.

Monday November 9
Partly cloudy with light air from NNW. Today being King Edward's birthday the flags were up on our vessel as well as the steamer. Tonight they are firing a salute of twenty-one guns and sending up sky rockets. One of our sleds came today from deer hunting. Developed a few negatives. One rabbit and a fox taken today.

Tuesday November 10
Wind moderate from NNE with a fine snow falling all day. Two rabbits taken today. The commander of the police, Major Moodie, has sent me a proclamation notifying me that after the first of January all vessels are to pay duty on all chargeable goods brought into the country, also forbidding the taking of musk-ox skins from the natives, or killing them. There is much more, but the main point is to prevent the destruction or

North-West Mounted Police at Fullerton Harbour, with additional administrative powers as 'Acting Commissioner of the unorganized Northeastern Territories.' He was never appointed 'governor' – as Comer states – but appears to have given this impression not only to Comer but also to newspaper reporters on his return south in the summer of 1904. To the *Halifax Herald* and the *Toronto Mail and Empire* Moodie was 'Governor of Hudson Bay.' When asked by Commissioner White of the RNWMP to explain the misinformation in the newspapers Moodie wrote, 'Regarding the statement that I possessed the authority of Governor of the Hudson Bay, I can only say that I know nothing of it, except that it has been frequently stated in various newspapers – such statement did not emanate from me' (Moodie to White, 30 August 1904, RCMP Records, vol 280, file 707).

extermination of the musk-ox, which I also think is proper. But I did not like to have this law affect me alone, and as he has not notified any others I protested against his discriminating against the American vessel.[13]

Wednesday November 11
Strong gale from NE with snow storm. Reading and writing is the way I am passing the time. I endeavour to write down all native customs and stories.[14]

Thursday November 12
We are having a perfect blizzard – wind NE and a driving snow storm. A person has to be close to the vessel in order to see it. Some of the natives came off to meals. Have been at work on buckets.

Friday November 13
Blowing very heavy during the night and this morning but moderating this afternoon, though still fresh and snowing. Wind NE. One of our men was

13 Comer felt that the Canadian measures were directed only against him but this was not true. Before arriving at Fullerton in the autumn of 1903 the *Neptune* had already announced the new regulations (appendix K (A) (iii)) at the Scottish and American whaling stations at Kekerten, Blacklead Island, and Cape Haven, in the Cumberland Sound – Frobisher Bay area, but there had not been time in the first summer to locate and notify the Scottish whaling vessels in Baffin Bay and Hudson Bay. In 1904, however, the Scottish ships *Active* and *Ernest William* were reached by a police patrol from Fullerton, and four out of the five other Scottish whalers operating in the Canadian Arctic were contacted by the *Neptune*.

In one respect, however, the restrictions did affect Comer more severely, at least temporarily. On arrival at Fullerton Moodie found that the trade in musk-ox skins was alarmingly high, threatening the survival of the herds west of Hudson Bay. He therefore prohibited the export of musk-ox skins on 8 November 1903, and made it an offence for non-natives to have skins in their possession. This restriction applied only to western Hudson Bay and was made known immediately to Captain Comer because he was close at hand. The fact that the prohibition was not announced at the same time to the Scottish whalers in the Repulse Bay region or to the American trader George Cleveland at Wager Bay was not indicative of discrimination against Comer, but simply a matter of logistics; they were a few hundred miles north and winter had closed in upon the ships at Fullerton Harbour. They would learn of the musk-ox regulation in due time.

14 Comer's observations on the material and intellectual culture, numbers, distribution and environment of the Eskimo inhabitants of northwestern Hudson Bay contributed significantly to the limited knowledge about these people. Aside from his own contributions to the literature, Comer's information and his artifact collections were utilized by anthropologists, notably Franz Boas (see bibliography).

quite sick with pleurisy last night. I doctored him by putting on his breast hot cloths with a little turpentine, which relieved him very much. Today is quite comfortable. Had Doctor Borden of the steamer prescribe for him, which he did by making up some medicine for him.

I am feeling considerably offended by Major Moodie's forbidding me to let my natives go out after musk-ox, while there has been nothing done to prevent other parties from hunting them or even trade for one. It looks very strongly like a slap at the American vessel, as we are the only one here. Common sense would say that due notice should be given to all and have the law go into effect at a fixed date ahead.[15]

Saturday November 14
Still blowing quite hard and driving the snow. Commenced banking the vessel. Wind NE. Major Moodie sent word to me that he would like to have my native men come to his house ashore, I presume to give them a talk as to the future. Got out beef and molasses.

Sunday November 15
Very pleasant, light wind NNE. I have lain awake so much thinking of the action of the police in regards to stopping us alone in trading, and the methods they have taken by having them come to his house and making them a speech telling them of the Great Father in Canada and making them a present each of a shirt or pants, then last night all were invited to a dance and supper on board the steamer.[16] In this way the new ad-

15 One can sympathize with Comer. In fact Moodie's official instructions (appendix K (A) (ii)) were to 'impress upon the captains of whaling and trading vessels, and the natives, the fact that *after considerable notice and warning* the laws will be enforced ...' (italics mine). But Moodie terminated the trade in musk-ox hides with no advance warning and announced his intention to charge duty on trade goods only after the *Era* was frozen in for the winter and unable to avoid the tax. When he notified the Eskimos on 14 November that they were subject to white men's laws and would be punished for violations, he meant from that very moment on. His actions appear to have been in direct violation of his orders.

16 Dr Borden referred to Moodie's speech as 'the powwow at Govt. House' and described it as follows:
   The Major had about eight gallons of tea made which with five pounds of hard tack and other biscuits soon disappeared. A clay pipe and a bit of tobacco was given to each of the twenty-five natives present. Then through the interpreter the Major told the natives that there was a big chief over them all who had many tribes of different colours and how this big chief, who was King Edward VII, had the welfare of all his peoples at heart. King Edward wanted them all to do what was right and good, and had sent the Major as his personal representative or in other

ministration was ushered in on a high tide. Feeling as though we were especially struck at, it has about made me sick. What will come next I can't say but expect they will try and drive us out of the country.

Monday November 16
Pleasant weather, light wind from west. Have partly banked the vessel over with snow.[17] Shall now wait awhile till the ice becomes stronger, as it would settle the ice should we put such a body on at once. The commander of the steamer sent word to me saying he would like to have my natives help bank his vessel, so they all went over this afternoon and worked. Took a couple of pictures while the men were at work. Got off a load of ice. Got our weekly paper, the *Neptune Satellite*.

Tuesday November 17
Pleasant, light breeze NNE. The men got off some ice. One man refusing to do as I wished, I put him in irons. When he was willing some two

---

words the King could not come himself so 'I was to act for him'! The Major wanted them to do what was right and good and to settle all quarrels but he would punish all offenders. (Borden, 'Memoirs,' 38)

Moodie then questioned the Eskimos about their practices of infanticide, parricide, and cannibalism, to which they understandably gave evasive answers. Following his interrogation Moodie 'with real ceremony presented a suit of woolen underclothes to each of the twenty-two adult Eskimo and a tuque, a pair of mittens and a sash to each of the boys' (ibid. 39). The irony of the situation, although not apparent to Moodie, was abundantly clear to Borden, Comer, and most certainly to the Eskimos. Here were a people who had maintained intimate ties with foreign whalers for more than forty years being treated as simple, helpless, credulous savages. Here were men who possessed whaleboats, darting guns, shoulder guns, and all the sophisticated paraphernalia employed in the pursuit of bowhead whales, who hunted with telescopes and powerful repeating rifles, and who normally wore American trousers, shirts, jackets, hats and sunglasses. Here were women who used manufactured domestic implements and containers, who made up clothes on sewing machines, who attended shipboard dances in imported dresses, and who bore children sired by the whalemen. To these people an officious, uniformed stranger was distributing underwear as if it were a priceless treasure and lecturing them on morals and their allegiance to a big white chief. When Moodie suggested that the Eskimos might wish to travel 500 miles to Churchill to send joyful messages of thanks to the king, no one responded.

17 Banking vessels with snow was standard procedure when vessels wintered in the Arctic. Wind-packed snow – the material of which igloos were made – was an excellent insulator and effective windbreak. It was readily available and easily cut into firm blocks. Snow was often packed on deck as well to restrict upward loss of heat from the hull.

hours later I released him. This is the first time I have ever done such a thing, but as the law forbids striking a man this is one way of subduing. One fox skin brought in.

Wednesday November 18
Pleasant with light air from north. One rabbit and one salmon got today. The banking which has been put up around the steamer was much more than the ice could stand around the bows, and the ice breaking, it went down out of sight.

Thursday November 19
Pleasant weather, moderate light breeze from NW. One rabbit and eight salmon brought in.

Friday November 20
Sky overcast, wind moderate from NW to north. Four salmon got today and two deer. There are many ponds within a few miles of us. Nearly all have salmon in (or salmon trout rather). The two deer were shot by our natives, who saw ten of them nearby.

Saturday November 21
Cloudy with a little snow falling and driving wind, moderate from north to NNE. The deer that were shot yesterday were brought in today, one being left at the vessel and the other taken to the natives' houses. Carried my letter over to the paper *(Neptune Satellite)* giving my views regarding the law which has come in force.

Sunday November 22
Partly cloudy, wind moderate from NW. One of our men seems to be subject to pleurisy. I was called up at 2 AM to attend him and put on hot cloths wrung out and a little turpentine sprinkled on them, which relieved him. I also used Iodine.

Monday November 23
Pleasant weather. A little cloudy. Light winds NW. Finished banking the vessel in and put cake of ice up outside against the windows. In my letter to the paper which is printed on the *Neptune* they looked it over and left out part of it which I wished very much to appear in regard to the law being unjust.

Tuesday November 24
Dull weather. Light wind NE, sometimes a little snow falling. Two foxes brought in today. One of them is probably a blue fox. Two rabbits also taken. Temp 3°.

Wednesday November 25
Stormy appearance. Wind hauling from NE to SE, light. Overcast. Two of our sleds went down to Depot Island yesterday for some walrus meat which they had cached there. They returned today. Temp −8°. Two rabbits taken today and three salmon.

Thursday November 26
Thanksgiving day. It has been cloudy with a moderate breeze from SE and east. We have had a little extra but no great effort. A dance is held on board the steamer but on account of the law prohibiting me from trading with the natives for musk-ox skins while others are not stopped has caused an unpleasant feeling, and I do not go to the steamer, and most of her people do not come here. Temp −13°.

Friday November 27
Overcast. Wind SE moderate. Snowing slightly. Temp 13°.

Saturday November 28
Wind moderate from SE to south with some snow. Having a dance here tonight. Temp 16°. These temperatures are at 8 AM.

Sunday November 29
Very pleasant weather. Light air NW to north, slightly cloudy. One fox brought in and one rabbit and four ducks. Temp air −4°.

Monday November 30
Very pleasant, light air from NW. Put out a cask to be used as a fox trap. One fox brought in today. Dealt out clothing and tobacco to the men. Opened a loaf of cake which was brought from home but found it had spoiled. Five months out today. Temp air −5°.

Tuesday December 1
Very pleasant weather. Moderate breeze from NW. Got up the small head of whalebone and split and scraped it. I believe, and so do the natives that caught it, that some of it has been taken. Temp air −10°. What we have of it weighed 244 lbs.

Wednesday December 2

Light breeze from SE with a little snow. Took a walk out to where we have a cask set as a fox trap about five miles off. A party of natives came – Santa Anna and Keckley. They are not my natives, though have worked for me before this. All natives of the Kenepetu tribe go to the steamer while all the Iwilic come here.[18] Air temp $-2°$.

Thursday December 3

Light breezes from north, quite pleasant. Today we cut the ice around the rudder and stem as the vessel is down low in the ice and shows signs of leaking more today, 840 strokes in the last twenty-four hours. They are having a dance on the steamer.

Friday December 4

Stormy appearance. Wind backing from west to north, moderate. One saddle of deer meat brought in. Mr Caldwell came over for a short time this afternoon from the steamer. Temp air $-18°$. The leak was 500 strokes today.

Saturday December 5

Strong breeze from NNE, varying in force, driving the snow. Am making another sled and having a chest made by the carpenter. Have had a dance on board this evening. Got out a barrel of pork and six cans of bread. Temp $-1°$.

Sunday December 6

Pleasant overhead but cold with a fresh breeze from NW which this afternoon is driving the snow along the surface. Temp air $-27°$.

Monday December 7

Slightly cloudy with a fresh breeze from NW, decreasing toward night. Am making a sled for a native. A party of Kenepetus came to the steamer, bringing three musk-ox and three wolverine skins and some meat. Temp $-22°$.

---

18 For the sake of convenience Comer and Low had agreed that the *Era* would have the services of able-bodied men of the Aivilingmiut and the *Neptune* those of the Kene-petu (Qaernermiut). There were approximately two dozen of the former and one dozen of the latter (Low, *The Cruise of the Neptune, 1903–4*, 27). In addition each vessel would have responsibility for the families of the 'ship's natives.'

Tuesday December 8
Fresh breeze from N by E. Partly cloudy. Took away the banking and cut the ice around the vessel and let her come up, which she did fourteen inches.[19] Commander Low of the steamer sent over word that he would like some primers and I sent him eight thousand, six of the No 1 and two of the No 2. Temp −14°.

Wednesday December 9
Moderate breeze from north to NW. Partly cloudy. Put the banking up again around the vessel. Cannot see as the leak decreases, as we got 450 strokes with the pump. Temp air −24°.

Thursday December 10
Wind moderate from north, a little fresh this morning. Temp −18°. Quite pleasant. A dance is being held on the steamer. In making out a list of the vessels and what they have taken since 1889 I find there have been seventeen voyages in all made here and have taken seventy-three whales in all, which have yielded about 77,775 lbs of bone. Of these the *Era* has so far taken twenty-nine of the whales which have yielded 33,800 lbs, and what we may take the remainder of this voyage I am in hopes will make her share much larger than at present. Temp air −18°.

Friday December 11
Moderate breeze from NE. Partly cloudy. Temperature air −14°. Our cook has been laid up and one of the boatsteerers has acted as cook today. Temp air −14°.

Saturday December 12
Fresh gale from NNE, driving the snow. Clouds broken overhead (temp air 8°). One of the men from the steamer was lost last night. They have been searching for him but can not find. I have also been out looking, can see but a short distance. Temp 8°.

Sunday December 13
Still stormy with the wind from NNE driving the surface snow along with blinding force. Temp 10°. They have not been able to find the lost man

19 The weight of snow banking erected around a wintering vessel depressed the ice surface and along with it the ship itself. Normally a whaler remained attached to the ice through the winter until the spring melt suddenly released it to rebound back a foot or so to its original level. When this happened the ship was said to 'rise out of

from the steamer. Today the sled left the steamer carrying letters to be forwarded to Fort Churchill. Over this I have felt as though the commander had not acted right. He may claim he sent me word but it was not direct from him to me, though he sent word to the mate and second mate – they have written and sent letters home. Very little value seems to be put on the several favors that I have done for them. Temp air 12°.

Monday December 14
Wind NNE to NE still, but not as strong today. Better weather to look around, but still unable to find the lost man. Temp air 6°. I had a very fine letter from Commander Low of the steamer thanking myself and crew for our kind assistance in searching for the lost man, also thanking for other favors. It made me feel much easier, for I have felt for some time that my endeavors were made lightly of. The schooner still continues to leak and if anything is on the increase, today being over 600 strokes.

Tuesday December 15
Partly cloudy, though a good day. We have all been off looking for the steamer's lost man. One of our natives thinks he saw the tracks of him well out to the floe edge. Moderate breeze NW. The schooner leaks more today – 1,000 strokes. Went over to the steamer this evening. Eight saddles of deer meat brought in today, some that had been taken last fall. Temp −8°.

Wednesday December 16
Moderate weather though overcast. Light wind from NW. Today we went out with a sled to where tracks had been seen of the lost man. We followed them in all their windings and circles but the general course was right before the wind, which took him to the edge of the ice about seven miles distant from the ship. He evidently kept walking out on the thin ice and went through and drowned.[20] Major Moodie and the interpreter went on

---

   her bed.' On this occasion, however, excessive depression of the ice surface early in the winter appears to have caused problems for the *Era*, so that remedial action had to be taken (see Comer's remarks on freezing a vessel in in appendix J).

20 The lost man, Frank O'Connell, described as 'a cabin boy of weak mind' (Low, *The Cruise of the Neptune, 1903–4)*, had left the steamer on 11 December to take a pair of skin boots to the native snow houses for repairs. When a snowstorm obscured visibility he evidently lost his bearings and missed the igloos. Describing the search Moodie wrote,
   We came to tracks about SW by S from steamer. There is no doubt about these tracks. Where we picked them up is about two and a half to three miles away. We

the sled with us. The leak was 625 strokes today. A child born. Anticoot among the natives. Temp air −4°. Nine ducks shot.

Thursday December 17
Light air from NW, sky overcast. They are having a dance on the steamer this evening. The leak today was 625 strokes. Opened a cask of bread and put in the locker. Temp air −2°. Some of the natives are building their snow houses near the vessel.[21]

Friday December 18
Blowing fresh from east with some snow. Have commenced building a small boat for the natives to use in sealing at the floe edge. Jack, one of our natives, has gone to Yellow Bluff to stay awhile. Temp 10°.

Saturday December 19
Wind moderate from east with a little snow. The men are having a dance on deck. The interpreter of the steamer does the fiddling. Temp 10°. Opened a new cask of flour. Some of the natives have moved off near the vessel. Two ducks brought in.

Sunday December 20
The wind was fresh during the night from NE. Today more moderate, some snow falling and drifting. Three bear skins were brought in today – had

---

found, close to an island that he had partly crossed, a few small round pellets of ice with some yellow wool sticking to them – as though he had pulled them from the collar of the sheepskin ... lined coat. From this he had run about 150 yards to the west, and then borne to the left, or before the wind which was NE or thereabouts towards midnight. His tracks led pretty straight from here, but occasionally he had wandered in a zig zag way always bearing to the left, as though he turned as soon he faced the wind. He had made several small circles also, in this way always to the left. These circles were quite small not over 7 or 10 paces across. The tracks led straight out to what had, at the time, been young ice which would not bear him, and there they ended abruptly. I have no doubt he went through at the first step ... We searched up and down the edge for half a mile or so on each side of end of trail, but saw no further signs ... The distance from steamer is between 6 and 7 miles (J.D. Moodie, 'Copy of Daily Diary,' 38, 39).

21 The annual snowfall in the central Canadian Arctic is low compared to many southern regions. There was usually not enough wind-packed snow for the construction of snow houses until late November or December. In the transitional period between summer tents and winter snow houses the Eskimos usually lived in temporary dwellings called quarmats in which a skin roof was supported by driftwood or whale rib rafters set on walls of stone, sod, or whale bones.

been killed last fall and the skins put under stone. Went over to the steamer a little while this evening but only saw the captain, the major remaining in his room while the commander was in the steerage. The vessel's leak is quite steady, 650 strokes each morning. Temp air 5°.

Monday December 21

Light airs with a little snow falling. Wind N by W. Mr Ellis has gone down toward Depot Island with a couple of natives hunting seal and ducks, to be gone three days. This afternoon met Commander Low, had a pleasant talk with him. He said that though they were here to watch what the American vessels were doing and to enforce Canadian laws he did not wish me to feel that there was any personal feeling against me. I told him I was, and had always been, sorry for not accepting his invitation to attend an entertainment which they held on the *Neptune* which he had sent me, but at the time I felt that they had picked out the American vessel to begin with on their laws.[22] He kindly sent over two pork hams, one for the men and one for the cabin, for a Christmas dinner. This may not be a regular article of diet Christmas days, but to us it will be a great treat for my people, as I am to come to the *Neptune* to dinner. While I have always meant well and tried to do well by the steamer people, and no doubt they may have done the same towards me, still we have become estranged till now. But Commander Low has better judgment than to allow such a feeling to exist and has acted to me very gentlemanly. Temp − 2°. One seal got today.

Tuesday December 22

Moderate storm from NNE, snowing. Wind increasing at night to fresh breeze. More of the natives are now building near the vessel. Mr Caldwell of the steamer (who edits the paper) came over a little while this afternoon, his object being to get some items for the paper. Commander A.P. Low, Mr Caldwell, and the Doctor Borden are the people who seem to feel friendly to us while all the others seem to avoid coming here and act as though they thought they were a superior people. I have an inward hope that they never will come, till they can come as *men*. The vessel continues to leak 650 strokes per day. Temp − 4°.

Wednesday December 23

Quite a pleasant day. Moderate breeze NW, then north. Temp − 18°. The natives got one seal at the floe and Mr Ellis came back from Walrus

22 See journal entry 26 November 1903.

Island, having been gone two nights with a couple of natives. They got three seal and two ducks. We are now making another boat. This will make five in use by our natives. Took a few pictures this afternoon of the new igloos and vessels. Temp − 18°.

Thursday December 24
Very pleasant with light winds north to west. Went over to the steamer *Neptune* where entertainment was given, Commander Low doing all in his power to make it a very pleasant Christmas Eve, assisted by Major Moodie and the other gentlemen of the ship. Dancing was kept up till 2 AM − but our time in the cabin not till 4 AM where a very liberal supply of all good things was set forth. Certainly no such a jolly time will be likely to be seen again in this harbor. A toast was drunk to the success of the schooner and my health. At the dance were a couple of the crew rigged up as Old Neptune and his wife and they acted their part exceedingly well. A couple of snapshots or flashlight pictures were taken. Temp air − 18°. During the day the Commander sent over three cans plums, two of pineapples, two of blueberries, two of cherries, three of pears, four of peaches, and one keg of gooseberry jam, for our Christmas dinner. The two large pork hams were sent over a few days before. It is not so much the goods but the man behind the act is what we appreciate.

Friday December 25
Fresh breeze and stormy appearance of the weather. A little snow. Took dinner on board the steamer. Had a fine dinner and plenty of other good cheer after drinking a toast to the king's health. Temp − 10°. One was given and drunk for the president of the United States. Another dance in the evening was held. The deck house was draped with many flags, including our flag, which was borrowed for the occasion, Mr Hearn, their mate, taking dinner on the schooner. I gave each of my men a pound of tobacco and the boatsteerers each some calico, also the mates.

Saturday December 26
Moderate weather, partly cloudy. Light breeze N to NNW. Each night the wind seems to increase, then moderate before morning. Finished another boat for the natives to hunt seal with. Sharpies is the name generally by which the boats are known. Temp air − 2°. Six seal and six ducks taken today. Had the graphophone out this evening. Our chronometer broke its hair spring (I think) and so now we are out of a valuable and needful article.

Sunday December 27
Pleasant weather. Light breeze from north. Four seal taken today at the floe edge, which is about six miles out from here. Mr Ellis has gone down to Walrus Island, near Depot, to stay a few days sealing – a sort of a pleasure trip. Mr Caldwell came over for a short time this afternoon. Temp air −10°.

Monday December 28
Very pleasant and clear. Have set up the blacksmith forge and run a lot of bullets – 2,500 for the natives' future use. Took a picture of the men while bringing off a load of ice. Light breeze north to west. Temp 20° below. The vessel continues to leak 750 strokes a day. Two seal taken today. Temp air −20°.

Tuesday December 29
Very pleasant. Moderate breeze from west. The natives got three seal today. I have been making a cane with a swan carved out of musk ox horn as a head for it. Temp air −1°.

Wednesday December 30
Cloudy at first, then clear with a moderate breeze from west to north. We are six months out today. The 30th of each month give the men what they need out of the slop chest. One of the natives came up from where Mr Ellis is camped sealing, bringing four ducks and one fox. They have not had any success in sealing. Our natives had a big anticoot meeting last night, in which they claim to have killed some evil spirit, so they do not work today but play games all day. Temp air −27°.

Thursday December 31
Quite a stormy day. Wind strong from NNE, driving the snow along. Several natives were intending to leave for Depot Island but did not start on account of the weather. The dance on the *Neptune* was also postponed. The vessel leaks about the same, a little over 700 strokes a day. Temp −17°.

3

The First Winter
(1 January – 10 May 1904)

Friday January 1
Weather improving during the night and today quite pleasant. Commander
Low and Major Moodie with Captain Bartlett come over for a short call,
also Mr Caldwell and Doctor Borden. Our natives got four seal and two
foxes. One of our natives went down to where Mr Ellis is camped. This
evening I went over to the steamer and spent the evening, returning at
twelve o'clock. A dance was held on board of her. Light breeze from
north-northwest. Temp air −23°.

Saturday January 2
Pleasant weather. Moderate breeze from north. Several of the gentlemen
from the steamer came over today, including Mr Ross and the Professor
(Halkett). The men came aft in a body tonight to complain of the soft
bread, it being quite heavy (but not unfit to eat), the men wishing to have
hardtack.[1] Gave the natives the soft tack and the men reprimand. Having

1 Dr Borden considered the quality and variety of the food on board the *Era* to be
somewhat less than satisfactory.
   The food was simply awful – salt horse, hard tack or ship's biscuits, black strap
molasses which I was told had been left over from the Spanish-American war and I
could well believe it. This food, already condemned, had been bought at a price
to suit the owners with no thought for those who needed it for survival. I have seen
a weevil at least two inches long when a hard tack was broken. One day I visited
a cache with one of the mates when he went to broach a tun of molasses to take a
supply back to the *Era*. The smell was so vile, that even in that wonderfully fresh
air, I had to run some distance to rid myself of that smell. I wondered how the crew
could ever swallow any ... (Lorris Elijah Borden, 'Memoirs of a Pioneer Doctor,'
65). Borden stated that there were no potatoes or canned fruits or vegetables on
board, but the *Era*'s provision list (appendix c) shows fresh and evaporated pota-

a dance tonight. The leak still continues at a little over 700 strokes per day. Two seal and two ducks. Temp air − 26°. Got the chronometer going again, could not see that anything was wrong.

Sunday January 3
Wind east but not bad weather. The natives have been out looking for walrus on the drift ice but saw none. One seal taken. Temp − 33°. These temperatures are all taken between 8 and 9 AM.

Monday January 4
Wind fresh from SE then backed to east, snowing moderately. Our natives were off again looking for walrus but did not find any signs of them. This place Cape Fullerton having become a port of entry on the first of January and having previously made arrangements to meet the collector, the Major Moodie, I went to his headquarters today and with a report and manifest of the cargo, the duties on my extra supplies amounting to ninety-four dollars and eighty-six cents, of which I signed an order on the owner.[2] Temp − 4°.

Tuesday January 5
Very pleasant with a light breeze from NW. Mr Ellis came back today bringing four seal, three foxes and four ducks. Our native got one seal. Temp of air − 24°.

Wednesday January 6
Moderate storm from east to SE with a little snow. Our natives have been out on the ice looking for walrus but saw none. Commander Low has been over a short while today and this evening I attended a lecture on

---

toes, onions, raisins, dried apples, dried and evaporated peaches, pears (probably canned), tomatoes, and corn. The quantity and quality may not have been satisfactory, however.

2 Because the whaling owners and masters had been given no advance notification that customs regulations were to be imposed, few were prepared for the new procedures. Staff Sergeant Dee reported,

I went to Repulse Bay – and found collecting customs up hill work as the Captains had no invoices, bills of laden [sic] or price lists [–] a small pass book with a list of articles jotted down, they claimed this was all they had for years as they were never called upon for Custom dues before. As I was unacquainted with the price of articles I took an open ——— from the Capt so as the authorities at Ottawa could fill in the amount and collect (Staff Sergeant Dee, 'Fullerton Diary 18 July to 17 Sept 04, '9. RCMP Records, vol 281, file 716).

the evolution of fishes given by Professor Halkett of the steamer. Two ducks brought in. Temp of the air −18°. Everyone from the other ship seems to be on a much more friendly footing now towards the schooner.

Thursday January 7
At first cloudy with a little snow with the wind NE but clearing up by night though the wind is NE but light. A dance on the *Neptune* this evening. One of our men is a little sick so that I have the doctor from the steamer come and attend him. Temp −16°.

Friday January 8
Moderate weather, light air E to S. A few clouds and hazy overhead.
Mr Ellis and a few natives started for Walrus Island to stay a few days sealing. Four of the steamer's natives with their families started off this morning hunting. Mr. Caldwell came over and spent the evening. Temp air −24°.

Saturday January 9
Moderate storm from SE with a little snow. A couple of our natives have gone to Walrus Island to join those who went yesterday – Stonewall and Gilbert with their families. Have had a dance this evening. Opened a cask of beef. I am having a number of ivory carvings made, also a few models of the natives' kayaks. Temp 0°.

Sunday January 10
Fresh breeze from east increasing towards night with a little snow. Our sick man is on the gain. The doctor comes to see him. Went over to the steamer this evening. Had a pleasant time. All seem to try and make it agreeable. Temp 1°.

Monday January 11
Wind fresh from SE with a little snow. Took a plaster cast of a man's face and hands – Takkie-li-Keeta (Sam). Opened a cask of meat. Temp air 6°. Had the graphophone going this evening and took two records of the natives' songs. I wished the old natives to go through the form of anti-cooting but the old ones could not look upon the idea with favor as it might offend their guardian spirits, but said it would be alright for some young chap to go ahead, which was done.

Tuesday January 12
Quite pleasant weather. Wind light from south. The natives went out to

the floe but did not see any seal. I got today a large granite-ware wash basin from the commander of the steamer to use in mixing up plaster [of] Paris in. Took a cast of a woman's face and hands – Mrs Cooper, wife of Sam, whose face I took a cast of yesterday. Temp air − 6°.

Wednesday January 13
Pleasant with a fresh breeze from WNW. Took plaster casts of two natives, Santa Anna and his wife Fatty – their hands and faces. Commander Low and the doctor and Mr King have each been over today. This evening I attended a lecture given by the commander, the subject being Hudson Bay and its explorers and resources. After the talk being over we retired to the after cabin where I was made a present by the commander of a fine pair of marine glasses, evidently expensive ones out of his personal property.

Thursday January 14
Very pleasant with fresh breeze from NW. A sled came up from Walrus Island for more provision. They had not taken any game but ducks and rabbits. Got off and opened a cask of bread, 759 pounds. One seal and two ducks got today. A dance is being held in the steamer. I have been over a while this evening. Temp air − 29°.

Friday January 15
Very pleasant. Light breeze from NW.. The natives have been to the floe but did not have any success. Had the doctor come over to see the man who is somewhat sick though better, his tongue being coated. Also one of our native women is quite sick (Nu-er-chin-nock). Temp − 31° below.

Saturday January 16
Moderate breeze NW, increasing and driving the snow. Mr Ellis and some of the natives returned tonight from Walrus Island bringing eight ducks. Two seal have been taken. Today I sent over to the steamer a box of magazines which had been sent to me by friends in Willimantic [Conn.] first taking out a pair of pants, a coat, and a vest which I had not known were in the box. The leak holds the same. Temp − 35°.

Sunday January 17
Pleasant with a fresh breeze from west, clear weather. I have got so that I look forward each week for our weekly paper, which is typewritten on the steamer. Temp − 36°.

Monday January 18
Pleasant but partly cloudy. Have been taking plaster casts of some native faces, five of them, as these are now getting ready to go to Wager River. Three seal were taken today. Wind west, hauling to NE. Temp − 30°.

Tuesday January 19
Very stormy. Wind NNE and north at night but more moderate. We have the best and most comfortable house over the after part of the vessel that we have ever had. Temp air − 20°. The vessel's leak is still 800 strokes per day.

Wednesday January 20
Very pleasant with light winds from NW. Two seal were taken today, also one duck and a few peterlarks. Attended a lecture given by Dr Borden on prevention of disease. My chief native Harry (Teseuke) has been making some ivory carvings for Commander Low, which he took over this evening. Temp − 39°. Stonewall came up from Walrus Island – no success in sealing.

Thursday January 21
A stormy day. Wind SSE strong with some snow. The natives have been out on the drift ice looking for walrus. They saw several but did not succeed in harpooning any. There are four natives at Walrus Island and I hope they may have been successful as oil and meat are very scarce in the igloos. A dance on board the steamer this evening. Temp − 20°.

Friday January 22
Strong breeze from east to NE with a little snow, clearing at night. One of the boatsteerers (Brass) is now feeling quite sick, I think mostly caused by the life he has led. The Doctor Borden came over today and gave him some medicine. Temp − 13°. Some of the natives are getting ready to go to Wager River sealing.[3]

Saturday January 23
Strong breeze from NE driving the snow. On this account our natives did not start. While out to one of the snow houses I saw one of the women

---

3 Comer meant Wager Bay, the deep indentation south of Repulse Bay. On maps of the eighteenth and nineteenth centuries it was variously called 'Wager Water,' 'The Wager,' 'Wager Inlet,' and 'Wager River.'

crying. She is the widow of the man who died last voyage – Paul – and seeing the men getting ready to go hunting brought to her mind when her husband used to go with the rest. The other natives took no notice of her but I sent and spoke to her telling her I mourned her husband's death with her. Temp −26°.

Sunday January 24
Very pleasant. Light wind NNE backing to NW. A party of fifteen [of] our natives started to go to Wager River to be gone till April. One seal and nine duck brought in. We have sixty-six natives in all who are maintained by the *Era*. Commander Low and Captain Bartlett came over and spent the evening. I was well pleased with their call. Temp −30°.

Monday January 25
Very pleasant with light winds from NNW. Took three pictures of the natives inside their igloos. The women have been working a number of white seal skins for me to be made up later. Temp air −30°. Eight hundred strokes is still what the vessel leaks.

Tuesday January 26
Pleasant with light winds from NW. Mr Ellis went with a party of natives to hunt seal and ducks at Walrus Island. One seal was brought in today. Took a plaster cast of our head native's face (Harry) (Teseuke). Temp −30°. Brought off two cans of powder and twenty-four cans of butter, also a bag of coffee, from the house ashore.

Wednesday January 27
Very pleasant with light wind from NW to west. No success today at the floe by our natives. A native came to the steamer today from Chesterfield reporting that there is much sickness among the people there. This evening attended a lecture given by Major Moodie on a trip overland to the Klondike. It was quite interesting. Had a very pleasant evening on board the steamer. Temp air −35°.

Thursday January 28
Cloudy with the wind WNW, a little snow falling. At night wind increasing. The natives did not go out to the floe today on account of the young ice which makes off. A dance on board the steamer this evening. Our man who is quite sick, though up part of the time, does not improve. Dr Borden does what he can for him. Temp −25°.

Friday January 29
Partly cloudy with the wind moderate from the west. One seal taken today.
Commander Low came over and made a call, a very pleasant one. Doctor
Borden also came over to see our sick man, who he seems to think may
have a touch of scurvy.[4] Temp the air −32°.

Saturday January 30
Pleasant with a fresh breeze from north. Mr Ellis came back today from
Walrus Island. They had killed one walrus but could save but part of it,
the ice being thin and breaking off. Also got three seal and two foxes
yesterday. Commander Low sent over some dried vegetables for the use
of the men who the doctor thinks have a touch of scurvy. Temp −39°.

Sunday January 31
Partly cloudy with a moderate breeze from north to NW. Took dinner on
board the steamer with Commander Low. Had a very pleasant time. My
head native has been doing some carving of ivory for both the commander
and Major Moodie and has done some splendid work. Temp air −32°.
Mr Ellis returned to Walrus Island with some natives, says he is going
to get a lot of seal, walrus and ducks. I opened a box of walnuts which
Mrs Balen sent, and ate them up.

Monday February 1
Strong breeze from NNE driving the snow. This evening had the graphophone out and the cabin full of natives. I took two flashlight pictures but
have not developed the plates yet. The first flashlight I probably did not
fix right as it came near taking the roof of the cabin off – at least the
report sounded loud enough to have done so.

Tuesday February 2
Moderate breeze from N by E backing to NW and becoming light. Doctor
Borden come over for a short time and brought over a bottle of liquid

---

4 Dr Borden diagnosed three cases of scurvy on the *Era* and attributed them to the lack
of antiscorbutics on board and failure to obtain sufficient walrus and whale meat
(Borden, 'Memoirs,' 70). Low was also critical of the absence of lime juice or other
antiscorbutics and agreed that the winter supply of fresh deer, seal, and walrus meat
was 'often very inadequate' (Low, *The Cruise of the Neptune*, *1903–4*, 266). Neither
Borden nor Low appears to have recognized that the presence of the *Neptune* and her
large crew may have been in part responsible for reducing the meat supply that
normally helped the *Era* keep her men healthy; it was Comer's complaint that the
government expedition contributed to scurvy on the *Era* (see journal entry 27 March
1904).

malt extract for our sick man. Although sick he is able to be up around. Temp air $-28°$. Had the natives in again this evening listening to the graphophone and got two records.

Wednesday February 3
Partly cloudy with a light breeze from NE by N to NW. Our natives got a large ground seal and one common seal. The large seal will make over a barrel of oil, which will be very useful to the natives as they were about out of oil in the snow houses. Temp air $-40°$. Spent the evening on board the steamer.

Thursday February 4
A stormy day, snowing. Wind NW to north, a fresh gale. According to native customs the natives (men) must not do any work for three days after killing a ground seal (they can go hunting). Neither may the women comb their hair for three days. This is to show their goddess that they appreciate the gift from her of the seal. Temp air $-16°$. Made a wooden funnel for running water into casks.

Friday February 5
Partly cloudy, wind NE moderate. The vessel now leaks 600 strokes a day. One seal taken today. I have been trying to repair a watch but made it much worse. A dance on the steamer this evening. Temp $-21°$. One of the small boats that we made the first part of the winter and use in picking up seal shot in the water has been carried off by the ice breaking farther in than the natives expected when they left it.

Saturday February 6
Partly cloudy. Wind NW to west, then to north. Sky becoming overcast and snowing, wind increasing. One seal taken. A dance this evening. This voyage we have not got a crew who can make up a good entertainment, not even a person to play the fiddle or the accordion, but we have to depend on the steamer. Temp air $-24°$.

Sunday February 7
Fresh breeze from NE during the day with a little snow, at night becoming clear and calm. Went over to the steamer a short while this evening. Temp air 2°. A child (Equark) died today, one of our natives. Sickness was probably brought on by the mother eating too freely of the large seal taken a few days ago. We made a coffin for it, which pleased the natives very much. It was then carried inland by three of the women. It was a sister

of a little boy who we call Tom Luce, named after our owner. This Tom Luce eats with me at the table sitting on my knee and receives much notice from all.

Monday February 8
Very pleasant with a moderate breeze from NE. The ice is now four feet thick. Two seal were taken today. Mr Caldwell was over this evening.

Tuesday February 9
Very pleasant with a light air from NE. Mr Ellis came back from Walrus Island. They had taken two large oujoug (ground seal) and two small seal and a few ducks, also two fox. The doctor and Mr Caldwell were over this afternoon. The vessel's leak now 600 strokes per day.

Wednesday February 10
Pleasant this morning but later becoming overcast and a strong breeze from west. Some natives went to Walrus Island, returning with those that came up yesterday. One salmon brought in. Temp air $-22°$. Went over to the steamer to hear Mr Caldwell give a discourse. Arrived late and only heard a part of it.

Thursday February 11
Pleasant but a fresh breeze from the west. Major Moodie came over with some documents, in connection with the duty on the imports, which he wished signed as they had not been made out at the time I went and entered at the custom house. We make a keg of spruce beer each week and it is generally very good. The doctor and Mr Caldwell were over today. The doctor brought over a can of beef extract and a can of tomatoes for the sick man. Temp air $-26°$.

Friday February 12
At first light air NW and overcast, then later hauling to NE and breezing up. The mate Mr Hearn and second mate Mr Bartlett were over and spent the evening. The commander and the doctor were over today. The sick man seems to improve a little under the doctor's care. Temp $-28°$.

Saturday February 13
Pleasant weather with the wind NNE moderate. A sled came up from Walrus Island, our natives there having taken one walrus and one oujoug (ground seal), also one common seal. Mr Caldwell was over a little while this afternoon. Temp of air $-30°$.

Sunday February 14
Very pleasant with light winds from north. One seal taken. A number of walrus were seen. Some natives returned to Walrus Island. Opened a valentine which had been sent from home by a lady friend, Miss Kate Gillette. Temp −35°. One seal taken.

Monday February 15
Very pleasant with light winds from west. Doctor Borden of the steamer is on the sick list. Temp of the air −36°.

Tuesday February 16
Very pleasant with light north winds. Today got several of the women to tattoo their faces with paint, as the tattooing on their faces will not take and show in a photograph. In this way I got five very good pictures showing as many different tribes – the Iwilic, the Netchilic, the Kenepetu, the Ponds Bay, and Southampton styles. Commander Low and Major Moodie also took pictures of the same. The commander came over this evening and gave me some good instruction in developing the plates. Temp of air −34°.

Wednesday February 17
Storming moderately with the wind SE moderate. The natives went off on the drift ice and got one walrus. The natives from the steamer also got one. This evening it came my turn to do the talking on the steamer for the entertainment. My subject was sealing and sea elephant voyages. Temp of air −18°.

Thursday February 18
Pleasant with a moderate wind but increasing in force north to NNW driving the snow, clear overhead. Commander Low came over with his camera and took some more pictures of the natives as the light is much better in our house on deck than theirs. A couple of natives come up from Walrus Island. They have taken one walrus there. Temp air −17°.

Friday February 19
A stormy day. Wind NW driving the snow along. Temp −40°. There is always work to be done, such as repairing guns or broken implements.

Saturday February 20
Good weather with a fresh breeze from NW. The two natives returned to Walrus Island. Mr Caldwell came over for a short visit bringing some

medicines for our sick man, who does not improve much if any. A dance this evening. Temp −36°.

Sunday February 21
Pleasant weather with wind west, moderate breeze. This evening spent at the steamer. The commander makes it pleasant to go there for a call, also Mr Caldwell and the doctor. Temp −36°.

Monday February 22
Cloudy during the morning but later clearing up and very pleasant. Wind west moderate. Our sick man is about the same in appearance but probably is really failing, as the only nourishment that he takes is beef tea. Temp −29°. Opened a cask of bread.

Tuesday February 23
Pleasant with light winds from west. A party of three women went this morning to fish for salmon at the pond near the head of the bay, about fifteen miles away. They are to be gone two or three nights. Temp −34°.

Wednesday February 24
Pleasant the latter part of the day though during the morning stormy wind, west, strong. The commander of the steamer gave a discourse on his explorations of Labrador and the mineral wealth of that country, after which a number were called upon to sing.[5] He saying that he would sing if I would, so I did so. Temp −35°.

Thursday February 25
Pleasant but a fresh breeze from NW. One seal taken today and one duck. The dance was held on board the *Era* tonight as the house over the steamer

5 Low's explorations of northern Quebec and his contributions to geographical knowledge are impressive.
In 1890, the interior of the Québec-Labrador peninsula was still virtually unexplored. Ten years later, the blank on the map had been filled: geology, physiography, climate, vegetation, fauna and inhabitants of the region were known, at least in their great lines, thanks to the remarkable amount of exploratory work done by the Canadian geologist A.P. Low. In thirteen seasons, Low sailed, canoed, dog-sledded and snowshoed some 10,000 miles in and around the peninsula. He gave the first accurate picture of the country, traced the outline of the Labrador Trough and gave the first description of the iron ores it contains. He surveyed the course of the principal rivers and a large portion of the northern and western coasts. His huge 1895 report is still considered the best work ever to have been printed about the

is too cold, it being quite open. Temp − 44°. The women last night returned from the salmon pond – no success.

Friday February 26
Pleasant this morning but later becoming overcast. Wind hauling from NW to north, moderate breeze. The natives struck a walrus but the line parted and the walrus escaped. A woman and a boy who came up from Walrus Island yesterday returned today. Had the graphophone out this evening. Temp − 42°.

Saturday February 27
Very pleasant with moderate breeze from west. Opened a new cask of flour and a barrel of pork. Had a dance this evening. I think that I shall have to give the men a talking to as a few men from the steamer are trying to run the dance to the exclusion of others. Temp − 40°.

Sunday February 28
Pleasant through the day with light winds from west, towards night becoming overcast. I took dinner on board the steamer, the commander and the doctor having been over and made a little call during the day. As both ships only have two meals a day dinner comes at 4 PM. Temp − 41°. Two seal taken today.

Monday February 29
Pleasant with light winds from west. Dealt out tobacco and clothing to the men. We look forward to the spring and as the sun is now quite bright and affects the eyes we shall soon have to wear colored glasses. Temp − 39°.

Tuesday March 1
Quite pleasant though hazy. Wind NNW to NNE moderate. The doctor and Mr Caldwell were over for a short while. The vessel leaks the same – 600 strokes per day. Opened a box of candied ginger root sent by Mrs Seymore. Temp. − 38°.

Wednesday March 2
Pleasant with a moderate breeze from NW. It is the coldest day we have had. A sled came up from Walrus Island. They have had fair success in

region (Fabien Caron, 'Albert Peter Low et l'Exploration du Québec-Labrador,' *Cahiers de geographie de Québec*, 9, 18 [1965]: 182).

catching seal. One boy who came up on the sled was badly frozen between the shoulders and both arms and wrists. [I] spent the evening on board the steamer. Temp −48°.

Thursday March 3
Today has been colder than yesterday on account of the strong gale blowing from NW. Cannot see a great distance. Our sick man is on the decline. Today he had a spell of vomiting up a very black liquid. His tongue has been quite black-coated for some time. I sent word to the doctor and he came over and gave him something to help his stomach. Driving snow with temp −44°. The lowest temp was −53° by the steamer's minimum thermometer last night.[6]

Friday March 4
Pleasant though a fresh breeze most of the time from NW. The doctor and Mr Caldwell were over awhile. Our sick man seems to be about the same. The main spring to the graphophone became broken and I took it over to the engineer of the steamer, thinking he would make a much better piece of work repairing it than I would. Temp −36°. made out a list of the different tribes of natives and the number in each tribe for Major Moodie.

Saturday March 5
Pleasant wind from the west moderate, breezing up stronger at night. A couple – man and wife – of natives came today from the south of Chesterfield Inlet, a river known as Ferguson River. They only had two dogs. No trade to speak of – one fox not skinned, two coats, five deer skins and several pairs mittens and stockings, and a dozen salmon. A story was started that they had brought several musk-ox skins, which seems to have put Major Moodie in nettles, as he sent for the man to come and see him and made inquiries regarding what he brought. Temp −34°.

Sunday March 6
Cloudy. Wind at first NW, then NE, then SE to S light, and at night snow falling. Spent the evening on the steamer. Had a very pleasant time. Temp −12°, at night 1°. Six hundred strokes is our daily leak.

6 On board the *Neptune* at Fullerton Harbour weather observations were taken five times a day by the three members of the scientific staff. Observations included air pressure and temperature, wind direction and velocity, cloud cover and (in summer) sea water temperature (Low, *The Cruise of the Neptune, 1903–4*, 29). This minimum of -53°F was the lowest temperature experienced during the winter of 1903–4 (see appendix E).

Monday March 7
A mild SE wind with a little snow falling. The natives went out hunting for walrus, saw a couple but did not get either but got a bear. This is the only one seen so far this winter though bear tracks have been seen. Got off a cask of bread for cabin use as the other one opened a few days ago was quite buggy. This we feed to the natives. Temp 4°. Took the graphophone down the forecastle this evening for the benefit of the sick man.

Tuesday March 8
A strong gale from NE to north, the wind having backed from the SSE during the night, driving the surface snow. As my books on the weather report have been used up I will try to keep the records in my journal after this. I have taken plaster casts of the faces of the two natives who came a few days ago. Eight AM barometer 30.15.[7] Att. ther. 62°. Temp −27°. Wind NE by N force 8 – a driving storm.

Wednesday March 9
Pleasant with a light breeze from NW. One seal taken. A sled came up from Walrus Island. They have taken one walrus and some seal. Eight AM bar. 30.51. Att. ther. 62°. Wind NW force 1, clear. Temp air −20°.

Thursday March 10
Very pleasant with light winds from NW. Commander Low and Doctor Borden, also Professor Halkett, were over today. One seal taken. The native who came from Ferguson River started on his return this morning. Eight AM bar. 30.77. Att. ther. 56°. Wind NW force 1, clear. Temp −26°.

Friday March 11
Very pleasant, light wind N by W to WNW. Some of the native women returned from salmon fishing, taken some thirty of them. They saw two deer and many tracks. Eight AM bar. 30.90. Att. ther. 67°. Temp air −20°. Wind N by W force 1, clear weather. At night light air WNW and a low

---

7 Comer's weather reports became more substantial at this point. To his usual observation of outside air temperature, wind direction, and wind strength, he adds barometric pressure (in inches of mercury) and temperature by 'att. ther.' or attached thermometer, used to apply corrections to the barometer reading. In addition, he begins to express wind velocity in 'force' from one to ten on the Beaufort scale (see appendix D). This increasing sophistication may have been the result of instruction by some of the *Neptune's* scientific personnel, most likely C.F. King, the meteorologist. As the barometer was inside the schooner's cabin the reading on its attached thermometer indicates the cabin temperature, here a comfortable 62°F in late winter.

bank of clouds in the west with light clouds higher up. Bar. 30.95 [at] 6
PM.

Saturday March 12

Very pleasant with light air N by W to NW, at night becoming overcast.
There is a wolf which keeps near the vessels though it has been shot at
several times. Today pulled out a couple of teeth for one of the men.[8] The
doctor and Mr Caldwell were over, also the professor. Eight AM bar.
30.86. Att. ther. 64°. Temp air − 16°. Wind N by W force 1. Blue sky 8
[tenths], a few clouds in NW.[9]

Sunday March 13

Very pleasant though a light haze. Some of our natives have gone to look
for deer and salmon. Mr Ellis going with them (wind NW force 2, hazy).
Three sleds came up from Chesterfield Inlet (Kenepetu tribe), two of them
going to the steamer, one stopping here. Thirty-four fox skins were brought
here but many more than that went to the steamer. I also got eleven deer
skins and a few pairs of stocking and slippers. Commander Low and
Doctor Borden took supper with us this afternoon. Eight AM bar. 30.77.
Att. ther. 68°. Wind NW force 2. Temp of air − 14°. Hazy at night, wind
increasing.

Monday March 14

Very pleasant with light air from NNW. The wolf that has been around
was shot today by one of the steamer's natives. Had the graphophone
out this evening for the benefit of the two natives who brought their trade

---

8 American whaling masters had to deal with a variety of medical problems. Equipped
   with a standard kit of medicines, surgical instruments, and a manual of instructions,
   they acted as general practitioners among their crews, doing what they could in any
   emergency or illness. Pulling teeth was routine and fairly straightforward compared to
   the amputations sometimes necessary in frostbite cases, and yet there was room for
   error, as Comer's experience indicates:
      Today our third mate Mr Silver complained of having the toothache and wished I
   should pull out the offending tooth. Certainly I could do that for him and after
   locating the tooth I applied the forceps and started pulling and soon had, as I
   thought, the tooth out. But like many things I do, I found I had pulled the wrong
   tooth − a perfectly sound one. I then told him I could surely get the right one next
   time if he would let me try again but he declined and is still going around with
   his tooth aching. (George Comer, Manuscript Journal on Board the *A.T. Gifford*,
   1910–12, G.W. Blunt White Library, Mystic Seaport, Mystic, Conn. 30 July 1910).
9 Comer now begins to record the cloud cover in tenths, a standard meteorological
   practice he appears to have learned from the scientific staff of the *Neptune*. His entry

here. Eight AM bar. 30.75. Att. ther. 53°. Temp air − 16°. Wind NNW force 2, clear.

Tuesday March 15
Very pleasant though partly cloudy. Wind west light. Mr Ellis and his party returned. They brought in twenty-three salmon, the largest one weighing 15½ pounds, but all are of fair size. The vessel is leaking much worse this morning, having to pump some 1,600 strokes. We cut the rudder and stem clear, hoping she will rise up out of her bed on the high tides of this new moon. Eight AM bar. 30.59 Att. ther. 58°. Wind W force 2. Temp air − 16°. Blue sky 4, clouds and haze 6.

Wednesday March 16
Light snow fell during the night and this forenoon with light air from NE. One of our natives came in who has been away for a few days. He has taken two deer and a few salmon. He went away mostly to get some soapstone but was unable to find it on account of the snow. Eight AM bar. 30.22. Att. ther. 65°. Temp air − 7°. Wind NE force 2. Hazy with a little snow falling.

Thursday March 17
Light airs from NE backing to west. Very pleasant. On both vessels the colors have been set, it being St Patrick's Day, and a football team match on the ice between the vessels. Eight AM bar. 30.27. Att. ther. 60°. Temp air − 20°. Wind force 1, clear.

Friday March 18
Very pleasant. Wind west light air, clear sky. Took away some of the snow forward of the waist both sides but the vessel does not rise out of her winter bed as yet. The leak has reduced to 1,000 strokes this morning. Mr Ellis and Mr Hearn of the steamer (mate) have gone away with a few natives to look for salmon. Eight AM bar. 29.93. Att. ther. 59°. Temp air − 22°. Wind west force 2, clear sky.

Saturday March 19
Pleasant though slightly cloudy. Wind light from west increasing in the evening. Two seal taken today. We cut the ice around the vessel and the schooner came up at 5 PM, it then being high water. She came up sixteen

---

means that eight-tenths of the sky was cloudless; by implication the clouds he observed in the northwest covered the remaining two-tenths of the sky.

inches.[10] A dance tonight, the leak today was 1,100 strokes. Eight AM bar. 29.80. Att. ther. 66°. Temp [−]24°. Wind west force 3, hazy. A white whale was seen today. This is the first one seen today [to date].

Sunday March 20
Pleasant weather. Wind moderate from west to WNW. A little hazy, at night clear. Took dinner on board the steamer and spent the evening there – a pleasant time. Eight AM bar. 29.76. Att. ther. 56°. Wind W by N [force] 5. Temp air − 30°. Slightly hazy.

Monday March 21
Pleasant weather. Moderate breeze from NW to NE. Hazy in the morning and again at night. The natives got four seal. Commander Low was over a short while. I am now making a new martingale for the schooner. Our carpenter is now working for Major Moodie on his house. Four seal taken. Eight AM bar. 30.26. Att. ther. 54°. Temp air − 24°. Wind NW force 1. Slightly hazy around the horizon.

Tuesday March 22
Slightly overcast with wind SE backing to NE moderate breeze. The natives got a walrus. The doctor who has been insane on the steamer during the winter is now reported to be very low and hardly expected to live. This doctor belonged to the police force. Commander Low sent over two boxes of dried apples. Eight AM bar. 30.10. Att. ther. 63°. Wind SE force 6. Temp of air − 1°. Thinly overcast.

Wednesday March 23
Partly cloudy but good weather. Wind moderate from NE One of our natives and family came in today from deer hunting, having only got two deer (Stonewall). The vessel's leak has now got down to her 600 strokes a day. We have still two sick men on the sick list. Both are being cared for by Doctor Borden of the steamer. Commenced on a new tierce of sugar. Eight AM bar. 29.94. Att. ther. 67°. Temp air − 11°. Wind NE force 4. Thick haze.

Thursday March 24
Pleasant with a moderate breeze from NE backing to NW and increasing to a strong breeze at night. Mr Ellis returned from his fishing trip, brought

10 See chap 2, n 19.

in five. Our cook has been for some time dissatisfied with his work and today requested to be allowed to move to the forecastle, which I granted, taking one of the men to fill his place. Our sick men do not improve. Eight AM bar. 30.14. Att. ther. 64°. Temp air − 20°. Wind NE force 5. Hazy.

Friday March 25
Pleasant weather with wind light hauling from NW to NE. One of my men giving me back talk and acting obstinate, I had him placed in irons. Then unbeknown to me two of the men made a complaint to Major Moodie on the steamer.[11] This I did not know till I got a note from the major wishing me to come over and see him. I declined to take notice of the note in the belief that he had no right to have anything to say in regard to the punishment of the men.[12] After I called the men together and found out who the men were who made the complaint I then confined them (five) in the hold. Then going over to the steamer I told the major what had been done, but after some talk (not cool) it was finally proved to my satisfaction that I was in the wrong in punishing a man to that extent of placing him in irons, but it was decided that on the whole I had better give the men a chance who made the complaint to withdraw it when I came back – the chance of doing so – or they would get worse treatment. The two who had made the complaint went over more than willingly and on their return

---

11 According to Moodie the two men had 'complained that one of the crew had been put in irons & put in the hold & asking if this was not a Canadian port where laws were in force' (J.D. Moodie, 'Copy of Daily Diary,' 58).
12 Moodie recorded that Comer 'was quite indignant, told me I had no jurisdiction on an American vessel, was going to hoist the flag and dared me to go on board' (ibid. 58). Although Moodie persuaded Comer that he had the power to intervene in a disciplinary matter on board an American vessel in a Canadian port if requested, Moodie was privately worried about the geographical and legal limits of his jurisdiction. If his authority did not extend into the District of Keewatin, as Moodie suspected, then 'anything I may do here therefore or anywhere on the west side of Hudson Bay is actually illegal ...' (Moodie to White, 9 December 1903, p 2 RCMP Records, vol 281, file 716). Moodie pressed the comptroller of the RNWMP for a precise definition of his powers, realizing that if his authority were only that of a magistrate he would not be able to try a serious case summarily unless he had the consent of the accused or the approval of the governor-general in Ottawa. In view of the length of time involved in obtaining consent from Ottawa or of sending accused and witnesses to a court down south, the law in the Arctic would be 'a dead letter' (ibid.). Later, after the incident with Comer, he recommended that a magistrate in the Arctic should be given the power to 'deal with cases where foreigners are concerned' (Moodie, 'Daily Diary,' 58).

I let all return to the forecastle, this being 1 AM this morning the 26th. Eight AM bar. 30.28. Att. ther. 68°. Temp −26°. Light air NW force 1. Thinly overcast, a little fine snow.

Saturday March 26
Pleasant with light wind from NE backing to NW. Have kept the men busy scraping snow off the house and down in the hold off the deck beams. Our sick men are not gaining. Old age seems to be against one of them while the other it looks as though the hard life he had lived was against his recovery. Doctor Borden is doing all he can for them. The steamer being here we have had less meat. Eight AM bar. 29.98. Att. ther. 53°. Temp air −28°. Wind NE force 4, clear.

Sunday March 27
Wind moderate, first north then hauling to SSE with light snow. The natives went out after walrus and got one, returning at 9 PM. This morning I called all hands in and made a few remarks regarding their late conduct and also gave them to understand that, having come on the voyage of their own free will, they had no one to blame, and the fact that the Canadian Government steamer being here and able to live in abundance of everything that money could buy, and all having clothes furnished them free, no doubt did make our means look small, even though we have sufficient of the necessaries of life. It was not my fault that the steamer is here. That she has aggravated the case there is not any doubt, though there has been no intention in it. The police force of whom Major Moodie is the head, whenever he could get a show, has not hesitated, in fact has been forward in showing his authority.

If the steamer had not been here we should not with doubt [doubtless] have had from the natives much of the fresh meat which has been traded to her. This of course would have done much to have warded off the strong symptoms of the scurvy, which is now keeping the men (two) sick.

I then told them that all liberty would be forbidden to leave the vessel without a permit, and I would also forbid the crew of the steamer and the police force from coming on board until a suitable apology was made for the amount of freedom used in giving back answers. One of the boat-steerers (Brass Lopes), who has been acting as third mate, was too free with his tongue to *me* and I returned him to the steerage. Later in the day I went over to the steamer and requested the Commander and Major Moodie to forbid their men coming to the schooner until my men showed a disposition to respect me more, which was soon shown on my return, and this evening I told them that everything from there would go on as usual. Six AM bar. 29.57. Att. ther. 58°. Temp −26°. Wind N force 1. Overcast, a little snow during day. Wind hauled to SE moderate, then in evening backed to north.

Sunday March 27 concluded
I then in the evening went over again to the steamer and reversed my request of the morning. There is no doubt but what our crew have imbibed a spirit of discontent from mingling with those of the steamer and police force who, as I have said, are extra well fed and well paid and clothed free, and while I can not blame the men for this feeling thus, as it is human nature, but still it brings an undue strain on my mind and would be as the point of a wedge, for a refusal to work and then would come mutiny, while at present there is not one among the crew who is in any way a bad man.

Commander Low, Mr Caldwell and Doctor Borden were over today. One of our natives (Stonewall) went to Walrus Island and brought up six boards which had been left there, also some seal skins which had been taken during the winter.

Monday March 28
Pleasant with light to moderate breeze from NW. Doctor Borden was over to see the sick men and took a drop of blood to examine from one of the men. He seems to think that they and even our carpenter show signs of scurvy. Eight AM bar. 29.39. Att. ther. 66°. Temp −21°. Wind NW force 4, clear.

Tuesday March 29
Very pleasant with light wind from WNW. Our thermometer got broken so that hereafter we shall not be able to keep a record of the temperature. Our two sick men were feeling quite a little improved today. Brass Lopes was allowed to take his place again in the cabin. Eight AM bar. 29.70. Att. ther. 55°. Temp air −27°. Wind WNW force 3, clear.

Wednesday March 30
Overcast lightly with wind moderate from SE. Commander Low and Captain Bartlett, Mr Caldwell and Doctor Borden were over, also the Professor and Mr Ross. Two sleds which went to Depot Island yesterday

morning after some boards left there by the *Francis Allyn* returned early this morning with heavy loads.[13] A wolf was shot by them and brought. Eight AM bar. 29.88. Att. ther. 64°. Temp air −14°. Wind ESE force 2. Lightly overcast.

Thursday March 31
Strong breeze at first from NE, backing to north and moderating. At night weather clearing as at first overcast with snow falling. Have made a new martingale and today fitted the iron work on. Major Moodie and Doctor Borden were over for a short time. Commander Low gave me a new thermometer to replace the one that I had and got broken. Doctor Borden brought over some medicine for me to act as a tonic, as he and the commander seem to think I have given too much thought to Major Moodie's interference with my affairs.[14] The major no doubt is a good honest man but his head is certainly swelled with his office and it is well known that seafaring men and military men never could get along well as a rule.[15] My box this month was from Mrs E.H. Chaffee, of fancy soaps – very welcome. Eight AM bar. 29.85. Att. ther. 61°. Wind NE force 7. Overcast, light snow. Temp air −13°.

Friday April 1
Very pleasant with light wind from north, at night hauling to NE and becoming hazy. The commander and Doctor Borden, also Mr Caldwell, were over. It being Good Friday the men had the day off and had a football match with the *Neptune's* men. Got beat. The vessel's leak is again in-

13 The *Francis Allyn* had wintered in Hudson Bay in 1901–2 then burned at sea in July 1902: her crew were delivered to Repulse Bay by the steamer *Active* and later taken home on the *Era* (Manuscript Journal of George Comer on Board the *Era* 1900–2, Old Dartmouth Historical Society, New Bedford, Mass).

14 On two occasions Moodie's exercise of authority upset Comer so much that Dr Borden had to prescribe medication. The first incident was the imposition of trade restrictions and customs duties on 10 November 1903, following which Comer had become ill. He later confessed to the doctor that he had suspected that his medicine was poison: 'I was afraid you was in league with the Major and trying to kill me.' Borden concluded that 'he seems to be slightly off' (Lorris Elijah Borden, 'The Lost Expedition, Being the Diary of L.E. Borden … 1903–4', 47). The second incident was Moodie's intervention in the disciplinary matter on the *Era*. The doctor observed that Comer seemed 'much worried and I believe it would not take a great deal to put him in a bad state.' Comer told Borden that he bore 'a fearful enmity for the Major' and would 'get even with him if ever the opportunity offers itself' (ibid. 76).

15 Dr Borden felt that Moodie 'rather considered himself a bit of a hero and was inclined to be somewhat boastful …' (Borden, 'Memoirs,' 36). Moodie's favourable self-image, as Comer suggests, may have been inflated.

creasing. Eight AM bar. 30.57. Att. ther. 64°. Wind N by W force 3, clear. Temp air − 25°.

Saturday April 2

Overcast, with a little snow and wind varying from SW to south and then SE. Two of the native men have gone to Walrus Island and intend to look for deer also. Had a dance this evening. One of sick men is getting better – the one who has been sick the longest. Eight AM bar. 30.21. Att. ther. 62°. Light air SW. Overcast, a little snow. Temp air 0°.

Sunday April 3

Overcast wind NE moderate, at night backing to NNW and increasing to a fresh gale. I took supper on the steamer and spent the evening. This dull weather seems to affect our sick men for the worse. Eight AM bar. 29.50. Att.ther. 62°. Wind NE [force] 2. Overcast, hazy with a heavy frost hanging from the rigging. Temp air 11°.

Monday April 4

Stormy wind NW, moderate gale. Doctor Borden came over again today. He cannot give much encouragement in the case of one of the men, the man's age being somewhat against him. A dance on the steamer this evening. Eight AM bar. 29.61. Att. ther. 64°. Wind NW force 8, driving snow, sun showing through. Temp air − 16°.

Tuesday April 5

Very pleasant with moderate breeze from WNW. Got four of the boats to the vessel to be got ready for spring work. Two of our natives came up from Walrus Island. They had taken three seal and one ground seal. Commander Low, Mr Caldwell, and Dr Borden were over. Our sick men remain the same. Eight AM bar. 30.26. Att. ther. 67°. Temp air − 13°. Wind WNW force 5, clear weather.

Wednesday April 6

This day has been marked with the first death which has happened on board since I have known and been in the vessel, which dates from 1889. This man is W.P. Maynes, who died of natural causes.[16] At first a cold

16 Dr Borden entered 'natural causes' on the death certificate but elaborated in his diary, 'There is no doubt but that the primary cause was Scurvy, as he had all the symptoms and in an aggravated form. The Scurvy was complicated by Bronchitis which ran on into a Catarrhal Pneumonia which carried him off' (ibid. 78). Moodie knew that scurvy

and then his age and general health were against him, being some fifty-five years of age. He had been shipped while under the influence of liquor. He had never been able to do his share of the work, such as going aloft or steering, but was always willing to do what he could. We had a funeral service at 2 PM, nearly all coming from the steamer to attend and also to the island where he was buried. Got a certificate of his death from Doctor [Borden] and had to present it to Major Moodie according to law. Eight AM bar. 30.40. Att. ther. 75°. Wind NE force 1, clear. The wind backed to NW and at night sky becoming overcast.

Thursday April 7
Quite pleasant with a light breeze from NW. We got off three more boats and have now commenced work upon them, also making new sails for some. Doctor Borden attends our sick man and not only furnishes all the medicines but the Commander has many things sent over that might taste good. This has been done all throughout men's sickness. Eight AM 30.24. Att. ther. 59°. Wind NW force 2. Light cumulus clouds, clear sky 4 tenths. Temp air 0°.

Friday April 8
Pleasant with light NW winds, at night hauling to NE and becoming overcast. Captain Bartlett and Commander Low were over. The Commander came over to take a picture of our crew for me and I have had a small frame made of the wood of the old martingale. We are making sails and fixing up our boats. Eight AM bar. 30.33. Att. ther. 66°. Wind WNW force 3, Temp air 0°. Light cumulus clouds, clear sky 4 tenths.

Saturday April 9
Cold damp NE wind, sky overcast. The proofs of the picture taken yes-

---

was largely responsible for the man's death and Comer undoubtedly knew it too, having recorded the presence of the disease on 27 March 1904. Dr Borden's description leaves little doubt as to the nature of the affliction:

Mien's [Maynes'] pulse very weak & fast. His Bronchitis is hard to control. His legs are stiff & swollen & has [sic] large spots on them which very much resemble black & blue spots. There are two or three of these larger than the palm of the hand. Most of the rest of the leg is covered with minute haemorrhages as large as pin points to 3 or 4 times that size into where the hair starts from. There are no hard nodules yet nor any effusion into joints. Gums swollen & very sore & spongy dirty red, breath very foul & has a nasty headache. Constipation. Seems pretty shaky today. Jake about same no haemolysis on legs yet, tongue black, breath very bad, gums much swollen & sore. Constipation severe (ibid 75).

terday are very fine. The commander brought over two. The doctor and Mr Caldwell also were over. The vessel is leaking about 1,000 strokes a day. Today I sold at auction the man's clothes who died on the 6th, the amount being $10.50. Eight AM bar. 30.36. Att. ther. 67°. Wind NE force 6, overcast. Temp air 1°.

Sunday April 10

Quite pleasant with wind moderate from NE, slightly cloudy. There is a sled with a party of natives and Mr Caldwell going to leave tomorrow for Wager River and Repulse Bay, so I have written two letters, one for Captain Murray and one for Mr Cleveland to send up by Mr Caldwell.[17] He is going on a prospecting trip. A sled arrived from the Chesterfield Inlet with some musk-ox skins for the steamer. Eight AM bar. 30.26. Att. ther. 66°. Wind ENE force 7. Air −1°. Cumulus clouds 5.

Monday April 11

Partly cloudy with light air from NNW. Mr Caldwell of the steamer started on his trip this morning. Today we commenced on three meals a day. The native who brought the trade to the steamer last night came over here with eight coats and three pair of pants, which I took in trade. Have sawed up a pair of whale jaws to be used as shoeing for the boats. Seven AM bar. 30.36. Att. ther. 70°. Wind NNW force 1. Overcast, a little fine snow falling. Air 2°.

Tuesday April 12

Very pleasant with light airs from NW. Two of our sleds that have been off deer hunting returned tonight with eleven deer – eight saddles and a carcass brought in to the schooner, the other parts being taken to the igloos. Commander Low and the doctor were over and the commander sent over a keg of mixed pickles which we prize greatly. Seven AM bar. 30.40 Att. ther. 67°. Wind NNE force 5, clear, wind becoming light during the day. Air −4°.

Wednesday April 13

Very pleasant. Light airs from north. All of our time is spent on boats.

---

17 J.W. Murray was in charge of the Scottish ketch *Ernest William*, wintering in Repulse Bay. Cleveland was at this time still an agent of the Thomas Luce firm (see chap 1, n 22).

The steamer people are surveying the harbor and neighboring islands.[18] The vessel's leak is now 800 strokes a day, has been recently 1,200. Seven AM bar. 30.43. Att. ther. 60°. Wind north force 1, clear. Air 0°.

Thursday April 14
Very pleasant with a light breeze from NW a little haze. One sled started off for deer hunting and also [to] bring in some that had been placed under stones. We are now shoeing the boats with bone from the jaws of a whale. Seven AM bar. 30.50. Att. ther. 67°. Wind NW [force] 3. Clear a little haze. Air −4°.

Friday April 15
Overcast with a moderate breeze from west and falling snow. The doctor comes over about every day to attend our sick man, who on the whole does not improve as he has some internal trouble. One of our sleds returned which has been gone some time deer hunting – have not had much success (Gilbert and Tom Nolyer). Seven AM bar. 30.32. Att. ther. 66°. Wind west force 4, overcast. Air 2°.

Saturday April 16
Strong winds from north, driving the surface snow. A dance was held this evening here. Commander Low and the doctor and also Professor Halkett were over. The commander brought over a fine lot of prints of pictures he has taken, also two plates, one of the *Era*'s crew and the other of crew and schooner. Seven AM bar. 29.54. Att. ther. 7l°. Wind N force 7. Cirrocumulus clouds 7. Air 12°.

Sunday April 17
Fresh breeze from north but becoming more moderate at night. Partly cloudy, snow drifting. Took supper on the steamer. Seven AM bar. 29.92. Att. ther. 70°. Wind north force 7. Thinly overcast. Air 2°.

Monday April 18
Slightly cloudy with light air from NW. We have got three boats finished

18 The survey of Fullerton Harbour was carried out in April 1904 by the expedition topographer C.F. King. He worked on the sea ice, taking several hundred soundings through holes bored by a machine contrived by the *Neptune*'s engineer, Crossman. The survey covered a radius of about thirty miles from the ship (Low, *The Cruise of the Neptune, 1903–4*, 29, 30).

with bone shoeing. Seven saddles of deer meat and eleven salmon brought in. The vessel's leak is now down to 600 strokes a day. Seven AM bar. 30.17. Att. ther. 65°. Wind NW force 4. Clouds cirrocumulus 9. Air 0°.

Tuesday April 19
Partly cloudy though pleasant. Out to work repairing boats. As we have ten of them to look after it will take some time. The two sleds returned having taken four deer. Seven AM bar. 30.22. Att. ther. 59°. Wind west force 3. Light clouds, amount 8. Air 2°.

Wednesday April 20
Partly cloudy with a fresh breeze from NE. Cold and raw. Seven AM bar. 30.07. Att. ther. 71°. Wind NE force 6. Thin clouds and a haze. Air 8°.

Thursday April 21
Overcast with light airs from NE hauling to SE. Commander Low and the doctor were over. I gave them some ivory carvings to take home to friends. Cold and raw. Our sick man is no better. One of the men does nothing else but look after him, which is great help to me. Seven AM bar. 30.31. Att. ther. 65°. Air −1°. Wind ENE force 4. Hazy.

Friday April 22
Overcast with light NE wind, at night clearing. This is my forty-sixth birthday. Seven AM bar. 30.10. Att. ther. 60°. Wind NE by E force 6. Overcast. Air 13°.

Saturday April 23
Pleasant with light breeze NE, backing to north. The vessel's leak is now 2,000 strokes. This is the most we have got in one day during the winter. Our sick man is some weaker today and there seems to be no hope of his recovery. The weather is quite cold and work on the boats is progressing slowly. A dance this evening. Seven AM bar. 30.35. Att. ther. 60°. Wind NE force 2, clear. Air 6°.

Sunday April 24
Overcast with a moderate breeze, first west then SW, then back to west, then at night NW. Took supper on the steamer and spent the evening. Our native women returned, having made up the deer skins into sleeping

bags.[19] I gave them out to the men. Seven AM bar. 30.42. Att. ther. 65°. Wind west force 3, overcast, Air 2°. (Three saddles brought in.)

Monday April 25
Partly cloudy with wind NE light. The weather is most too cold to work at putting new plank in the old boats which need it, so our work does not progress as fast as I should like, though we find plenty to do. The vessel's leak today got down to 900 strokes. Seven AM bar. 30.30. Att. ther. 71°. Wind NE force 6, clear. Air 9°. By Commander Low's observations the latitude of this place is 64° 00'N but is laid down on the charts as 64° 05'N. It is also about fifteen miles too far east by the charts, the correct longitude being 89°16'W while the chart claims 88°44'W.[20]

Tuesday April 26
Overcast with wind E to SE strong breeze, at night ESE. Busy on boat gear. Seven AM bar. 30.27. Att. ther. 73°. Wind east force 4, overcast. Air 17°.

Wednesday April 27
Overcast with light winds from east. At night light fog. The doctor who has been sick and out of his head all winter on the steamer died this morning and was buried this evening at 7 PM. As it is impossible to dig a grave in this country the coffin is surrounded and covered with stone, the colors being hoisted at half-mast on both vessels. Spent the evening on the steamer. Seven AM bar. 30.00 Att. ther. 67°. Wind ESE force 7, overcast. Air 13°.

Thursday April 28
Overcast at times, then partly clear with a strong breeze from west to NW. Cleared the snow off from the deck and worked some on the boats.

19 The caribou skin sleeping bags were made up for the men who would be leaving on the spring whaleboat cruise on 12 May.
20 Contemporary charts of Hudson Bay contained a number of inaccuracies. For example, the fact that a strait existed between Southampton and Coats islands, providing a direct route to the whaling grounds, was discovered by American whalemen in 1864, four years after the initiation of whaling in the region. Fisher Strait, as it came to be called, was not incorporated on the Admiralty charts, however, for another eight years. Again, the eastern extremity of Southampton Island, named Bell Island on some charts and maps, proved to be a peninsula of Southampton Island in 1900, when Captain Comer attempted to sail through the non-existent passage. In addition, known coastlines and anchorages were often placed at the wrong longitudes, owing to the unreliable timekeeping and navigation of the earlier expeditions that had first laid down the outlines of the land.

Commander Low and the doctor were over for a short visit. The snow bunting are quite plentiful around the vessel. Seven AM bar. 29.64. Att. ther. 76°. Air 28°. Wind wsw force 4, overcast. Some of the natives are now moving into the tents.

Friday April 29
Mostly overcast with moderate breeze hauling from north to NE. Took down the porch or outside doors so as to be able to take one of the boats into the house, which we did, as the weather is too cold and raw to work on them out on the ice. The first partridges were seen today, for this spring.[21] Seven AM bar. 30.19. Att. ther. 60°. Wind north force 1. Clouds 7. Air 14°.

Saturday April 30
Partly cloudy with a fresh breeze from east to east-southeast. The snow is becoming softer around the vessel. Seven AM bar. 30.11. Att. ther. 64°. Wind E by s force 6. Thinly overcast and hazy. Air 20°.

Sunday May 1
Partly cloudy with fresh breeze from NW. At night overcast. Four partridges were shot today but only one skin was fit to save. Spent the evening on the steamer. Seven AM bar. 30.08. Att. ther. 62°. Wind NW force 4. Clear with a little haze. Air 20°.

Monday May 2
Squally with snow, later clearing and a little colder. Finished one boat and put out, and took another in the house. Took off the main hatches for the first time this spring. Seven AM bar. 30.25. Att. ther. 69°. Air 21°. Wind west by south force 3, overcast.

Tuesday May 3
Clear but with fresh breeze from NW driving the snow a little and quite cold. Finished another boat and took in another. One of the sleds returned which went off during the winter to Wager River. They brought in ten

---

21 Comer is referring to ptarmigan, year-round arctic residents. Slow on the ground, the birds are easy to approach. Borden reported that as the increasing sunlight of late winter tempted the *Neptune* crew outside for longer periods, 'Some took long walks, others visited the trap lines, while others knocked over ptarmigans. These birds were so tame that actually knocked over is the right expression. However Commander Low had to issue an order finally restricting this sport as he was well aware of the need of conservation' (Borden, 'Memoirs,' 84).

saddles of deer meat (Jimmy and George). Their trip has amounted to nothing. The other sled has yet to be heard from. Seven AM bar. 30.26. Att. ther. 56°. Wind NW force 6. Clear. Air 1°.

Wednesday May 4

Partly cloudy with fresh breeze from NW. This morning Commander Low started off on a surveying trip to be gone about two weeks to work from Chesterfield Inlet up back here. One of my sleds went with him to carry a load as far as Depot Island for him. This evening the natives who have been gone during the winter looking for game returned. Our natives Billy and (Pikey's) Jimmy brought in three musk-ox heads, two bears, one wolf and one wolverine. The steamer's natives also returned but they had quite a few musk-ox skins. They have been to a different part of the country. We are making quite extensive repairs on our old boats but getting along very well and I am much pleased with our carpenter, who in every way is proving himself to be a very likely young man. Seven AM bar. 30.33. Att. ther. 61°. Wind NW [force] 2. Clear sky. Air 5°. Twelve salmon brought in by Stonewall. Opened a barrel of pork.

Thursday May 5

Partly cloudy but mild light air hauling from west to NE and east. Put out another boat and took in one. Major Moodie was over and gave me a permit to own and export the three musk-ox skins and heads.[22] The sled which took a part of Commander Low's goods (stores) and himself returned this evening. Three seal were taken today – these are the first that have been shot while basking in the sun on the ice. Twenty-nine years ago today I started out on my first voyage whaling for this country in the bark *Nile*, Captain Spicer. Seven AM bar. 30.36. Att. ther. 70°. Wind west force 1. Clouds 3. Air 15°.

Friday May 6

The wind has increased to a gale from the east with snow. One of our natives has been helping Major Moodie land some coal this forenoon. We had musk-ox meat for dinner and think it superior to deer meat. Opened a cask of bread. Seven AM bar. 30.11. Att. ther. 64°. Wind east force 6, overcast. Air 19°.

---

22 Though it was unlawful to export musk-ox skins (see journal entry 10 November 1903), Moodie allowed Comer a special dispensation because of his contract to collect scientific specimens for the American Museum of Natural History.

Saturday May 7
Very stormy. Wind E by N, a driving snow storm. We have got the last one of the boats nearly completed in repairing. Seven AM bar 29.64. Att. ther. 68°. Wind E by N force 8. Overcast, driving snow storm. Air 20°.

Sunday May 8
Very stormy. Wind backing from NE to north, at night more moderate. Our sick man seems to be a little better. Captain Bartlett and Doctor Borden were over this evening. Air —. Seven AM bar. 29.82. Att. ther. 65°. Wind NE force 8. Overcast, snowing. Air —.

Monday May 9
The wind moderated a little during the night and backed to NW but has increased again to a gale, snowing and driving the snow. Got out the last boat from the house but we still have much more work to do on some of them in small repairs. Gave out the skin boots to the men. Got the three musk-ox heads skinned and cleaned and worked the two bear skins. Seven AM bar. 29.77. Att. ther. 66°. Wind NNW force 5. Overcast thin overhead. Air 14°.

Tuesday May 10
Partly cloudy, wind strong from WNW driving the snow. Such a body of snow has accumulated around the vessel that it caused the ice to settle over a foot. The natives have had to move out and make new homes or erect their tents as the water came in to their old houses. Seven AM bar. 29.93. Att. ther. 58°. Wind west by north force 6. Clear, snow driving. Air 7°.

4

Spring and Summer Whaling (11 May – 25 September 1904)

Wednesday May 11
Cloudy with a very little snow falling, moderate breeze from west. Dealt
out clothing to the men as we expect to start for the floe tomorrow, though
I shall remain with my boat's crew and do work which is needed to be
done.[1] Hope to locate the leak in the schooner and try and stop it. Had
a dance this evening. Seven AM bar. 30.35. Att. ther. 61°. Wind west force
4, overcast. Air 10°.

Thursday May 12
Took three of the boats out to the floe and left Mr Ellis and Mr Reynolds,
also a native crew in my boat. We have never had the floe edge so good
in other years as it is now. We shoved off and took a sail to the north.
Got a ground seal. Overcast at first, then clearing off, pleasant. Seven
AM no observations taken.

---

1 Whaling in May and June was carried out from the floe edge, several miles southeast
of the anchorage at Fullerton Harbour. By 1903 Comer had modified the operation.
Instead of camping at or near the floe edge to await the arrival of migrating whales
(as native whale hunters still do in northwestern Alaska) he and his men undertook
long cruises from the ice edge, during which the crews lived in their small boats,
hauling them out each night to sleep on board under special covers. On these cruises
the whalemen had to contend with pack ice, cold weather, fog, itinerant polar bears,
and other hazards. But Comer was acknowledged master of this method, often leading
his small fleet of half a dozen whaleboats up and down the mainland coast and even
across Roes Welcome Sound to Southampton Island, pursuing whales more than
100 miles from the vessel. It was impractical for the boats to transport bulky blubber
long distances to the ship for rendering into oil, but the high price of whalebone
made it possible to ignore the oil altogether. When the men made a kill they simply
extracted the bone from the whale's mouth and left the rest of the carcass.

Friday May 13
Pleasant though cold with a moderate breeze from NW. Our natives have all moved down to the point of land which forms Cape Fullerton. We have been getting lines ready for the three native boats and getting other work along. Commander Low returned from his trip to Chesterfield Inlet.

Saturday May 14
Pleasant though partly cloudy. The natives (men) came up this morning and took the three boats of theirs. We still have four left. Three of these will be carried down later. We are at work on the last boat, the seventh one. These with our three new boats make quite a fleet. Commander Low was over. The doctor comes every day. Our sick man is feeling better though he has some internal complaint. Moderate breeze west.

Sunday May 15
Very pleasant with moderate wind from NW hauling to north. Some of the women came up from the camp (now at the point of land which forms the entrance to the harbor). and got the remainder of their goods. Two more ground seal have been taken. Took supper on the steamer and spent the evening. Had Doctor Borden examine me today to see if I had any heart trouble. He said I had nothing wrong with my heart and was in good health.

Monday May 16
Pleasant though becoming overcast towards night. Wind light from ENE. Working on one of the boats and clearing away snow from the side of the vessel. The vessel's leak is now 1,200 strokes a day.

Tuesday May 17
Overcast with wind hauling from east to south and at night becoming foggy, which freezes. Three barrels of seal blubber were brought up from the camp. Today I gave the doctor for his services this winter two musk-ox skins and one bear, also two pup or cub skins.[2] In looking over our

2 Moodie's prohibition of trade in musk-ox skins affected *Neptune* personnel as well as those of the *Era*, and some were clearly disappointed that they could not purchase a few handsome skins as souvenirs of their arctic voyage. Borden was one; after Moodie's proclamation he wrote in his diary 'I am going to try and get a musk-ox skin or two out of the racket some way' (Lorris Elijah Borden, 'The Lost Expedition, being the Diary of L.E. Borden ... 1903–4,' 50). Comer had been allowed to retain the skins on hand at the time the new regulation was imposed, and from this stock a couple of musk-ox skins came discreetly into the possession of the doctor.

furs we have 101 musk-ox, 26 wolf, 15 bear, 96 fox and 4 wolverine skins.

Wednesday May 18
Wind hauling from south to west and clearing, coming off colder. Took
the coal out of the main hatch and put it in forepeak. We can locate one
leak near the stern part but so low down that we shall not be able to get
at it. There was a little fog last night. Some of the women came up today
and washed out the bear skins.

Thursday May 19
Wind north hauling to NE, light. Have been coopering casks and stowing
the hold. A gull was reported to have been seen yesterday, the first one
of the season this year. (Overcast.)

Friday May 20
Wind hauling from NE around to south and southwest. Have been stowing
casks and getting the vessel in order. Cloudy.

Saturday May 21
Foggy, misty weather. We got off from the island spare oars, boats' masts
and some cordwood, also crow's nest and topgallant yard. Wind moderate
from SW, a little rain. Saw a gull and a hawk.

Sunday May 22
Wind SW, hauling to west during the day. Last night rain and sleet which
coated everything with ice. This afternoon wind increasing and at night
a little snow. Took supper on the steamer and spent the evening. Today
I wrote a letter of thanks to the commander regarding the doctor, who
has been very kind in coming over and attending our sick.

Monday May 23
Cloudy with increasing winds varying from NNE to north and NW. Most
of the native men came up this morning and have been making bullets –
made 3,300. Got off our wood and two loads of ice. Now that this place
is to be occupied by the police we are thinking of wintering in some other
place next winter.[3] Commander Low gave us a large tierce of lime juice.

3 In June Moodie had read to Comer 'the law, under Customs Act, regarding going into
  harbor other than "ports of entry," ' informing him that by law he must not winter
  at any place other than Fullerton Harbour, the sole port of entry for the Hudson Bay
  region (Moodie, 'Copy of Daily Diary,' 73). There is no evidence that Comer's idea

Though we have no use for such a quantity we had nothing small to put it in so took the whole. Canadian vessels all carry large amounts of it and are generally known as lime juicers. The natives when they returned took two boats with them.

Tuesday May 24
Partly cloudy with strong breezes from west to sw. Doing small jobs in the way of picking up and getting ready for a change. This being the late queen of England's birthday, the colors have been up till noon when, as the gale was so strong, they were taken down.

Wednesday May 25
Snowing all day with the wind sw moderate. Working on small jobs. The vessel's leak is from 500 to 600 strokes a day now.

Thursday May 26
Snowing all day. We brought off all our butter and potatoes which we had stored in the house ashore. A sled came up from the boats and wanted bread, flour and molasses, also butter, which we sent. Had to open a cask of bread. Wind sw moderate.

Friday May 27
Overcast and some snow falling. Got up the squaresail yard and topgallant yard, also main gaff, and brought off some small stores from the house. Wind south SE and NE moderate. A sled came up and brought up $1^1/_2$ saddles deer meat.

---

of spending the winter at another locality was motivated by a desire to circumvent the regulations on trade and customs; it seems more likely that he simply wanted to get away from Moodie, whom he disliked intensely.

It is interesting that Moodie apparently did not insist that the Scottish ketch *Ernest William* must winter at Fullerton; she wintered in the region from 1904 until 1910, but never at Fullerton Harbour. The *Queen Bess*, tender for the Scottish mica mining station at Lake Harbour, left for Dundee in August 1904 without clearing through Fullerton Harbour (*Queen Bess*. Abstract logbook 1904. General Register and Record Office of Shipping and Seamen, Hayes, London). Furthermore, when the Dundee steamer *Active* returned to Hudson Bay in 1904, and again in 1905, she did not enter or clear at Fullerton Harbour, but simply stopped at Lake Harbour, proceeded directly to Repulse Bay, and reversed the procedure on the way home (*Active*. Abstract logbook 1904. General Register and Record Office of Shipping and Seamen, Hayes, London; *Active*. Ice log 1905. Danske Meteorologiske Institut, Charlottenlund, Denmark).

Saturday May 28
Overcast and snowing part of the time. Wind moderate and varying from
SE to NE. Bent the mainsail to the gaff and mast hoops. Commander Low
was over this afternoon. I spoke to him about taking our sick man and
transferring him to the supply steamer when he should leave here but he
thought it could not be done very well. I did not expect that he would be
willing to take him but to please the man was the reason I asked. A native
from the south of Chesterfield arrived today. He has walked up here. His
object in coming is to get a wife.

Sunday May 29
Partly cloudy though quite warm and pleasant. Took a picture of the two
graves and the major's houses.[4] Tomorrow intending to start for the floe
to stay with my boat's crew. Took supper on the steamer and this evening
had the graphophone out and got three records of native singing. Light
airs, west.

Monday May 30
Left the vessel at 7:30 AM to go to the floe edge. Got our boat down to
the water with the help of the native women and dog sleds. In the afternoon
all the boats returned from cruising. On the way down we met two of the
women taking two barrels of blubber to the schooner. Partly cloudy with
light variable winds.

Tuesday May 31
Cloudy with a little snow falling at first, later weather improving. Started
with two of the native boats with me (Harry and Ben's) to go to Cape
Jalabert, the south side of Chesterfield Inlet.[5] Got down as far as Walrus
Island. Saw quite a number of them in the ice, got one. Wind SE to NE
light. Very little ice offshore. We have never seen the bay [Hudson Bay]
so clear of ice at this time of the year.

4 The Fullerton police post consisted of living quarters measuring 15 by 24 feet and
   divided into three rooms, a storehouse for provisions, a coal shed, and a lean-to
   kitchen 12 by 16 feet with a large porch. There were plans to erect a 'barrack room,
   quarter master's and Trading Store' in the autumn of 1904 (J.D. Moodie, 'Report
   of Superintendent J.D. Moodie on Service in Hudson Bay ... 1903–4,' 5).
5 The name Jalabert was given to a bay not far south of Wager Bay on some eighteenth
   century maps, but by 1820 it denoted the land between Chesterfield Inlet and Rankin
   Inlet, which appears to agree with Comer's description. The name has no official
   status today.

Wednesday June 1
Snowing and blowing so that we are not able to proceed. Hauled the boats up higher. Wind NE. We are very comfortable in our boats. Put the walrus meat and two seal near the shore to be taken on our return.

Thursday June 2
We are now at Promise Island, Chesterfield Inlet. On our way down we stopped at Depot Island, where we had dinner, and got here at 9 PM. We left six more seal on an island about five miles north of Depot. Seven seals taken today. The ice is gone in the main channel up the Chesterfield Inlet as far as we can see. Winds north to northwest fresh.

Friday June 3
Overcast with a very little snow. Wind moderate from NW, at night calm. We left Promise Island and came to Cape Jalabert. Quite a number of deer were seen on the Cape. The natives went hunting and got eighteen which we helped them bring to the boats. Two seals also taken. We are hauled out on the shore ice. A little misty at night.

Saturday June 4
This is the place where whales were quite plentiful in years past, but not seeing anything favorable we started on our return trip and came up to about fifteen miles below Depot Island, where we hauled out on the edge of the floe. Wind light from E to south. My eyes are quite painful from snow blindness. One ground seal taken.

Sunday June 5 Depot Island
We started this morning at 4 AM with the wind SE and increasing in force, with snow. After four hours' run we made out to find a landing place at Depot Island, the ice being gone close to the outside of the island. The wind increasing to a fresh gale have had to haul the boats well up three times on the ice over the reefs. At night wind dropping off, misty weather.

Monday June 6
We are now six miles north of Depot Island. We were unable to leave the island till 2 PM on account of the ice drift. The seal we had buried in the snow near here were carried off with the ice as it had broken up considerably. The walrus is probably gone also. Twelve seal were taken today, also one ground seal. Wind at first NE fresh, later calm.

Tuesday June 7
Blowing a strong gale from SE and at night fog and rain. We can do
nothing but eat and sleep. The loose ice from the recent gale is packed
in and protects the ice we are on from being broken up, the open water
being a mile off. Three seal taken. We can do nothing but eat and sleep.

Wednesday June 8
We are about twenty miles from Cape Fullerton, not being able to come
far today on account of the loose ice. We found ourselves adrift at 1 AM,
the swell coming in and breaking the ice up fast, in to Winchester Inlet.
We met a sled coming down to us to see if we needed anything. Loaded
it up and started it back. Had a note from Mr Ellis. Wind moderate from
west. Weather variable, some snow. Seven seal taken and four geese.

Thursday June 9 Cape Fullerton
We started at 4 AM and came through quite a lot of loose ice. In one place
we had to haul over it. Got to Cape Fullerton at 11 AM. Wind south and
backing to SE fresh. The ice here has not broken up, nothing to be compared to that farther south. Here we can see but little change. Had a
wash and shave, which makes a great difference in one's looks and feel-
ings. Have quite a cold and sore throat. Found the other boats alright.
They have been painted.

Friday June 10 On board the schooner
Came up this morning to prepare for a start of four boats to go to Southampton to look for whales. The wind being east I did not return when
the sleds went back but sent all the provision needed. Commander Low
and Doctor Borden are going to take two of our boats over, for which he
[Low] will allow me the amount of $25.00 in canned goods.[6] He has already
sent over pears (one case), peaches (one case), raspberries (two cases),
blueberries (two cases), dried apples (two cases), pickles (two cases),
potted meat (one case), keg of pickles (one), sauce (one dozen bottles),
lime juice (one tierce), kerosene oil (two tierces), pepper (five pounds).

6 Comer had invited Low to accompany his whaling crews on this arduous and somewhat hazardous voyage to Southampton Island (A.P. Low, *The Cruise of the Neptune,
   1903–4*, 31). Low accepted, Borden tells us, in order to take formal possession of the
   island and make geological, zoological, and botanical observations (Lorris Elijah
   Borden, 'Memoirs of a Pioneer Doctor,' 88). Low in turn invited Borden, who in addition to his medical responsibilities was official botanist of the *Neptune* expedition.

Saturday June 11
A stormy day. Snow and rain with strong breeze from E to ENE. Commander Low, Mr Caldwell and Doctor Borden were over today for a visit, making the time pass pleasantly. My cold and sore throat seem a little better.

Sunday June 12
The wind being from the NE and blowing a gale with snow and rain, I have made no attempt to go back to the boats. Took supper on the steamer and spent the evening. Had a pleasant time. My cold is a little better.

Monday June 13
Still blowing a gale from the NE with snow and rain, some hail. Everything covered with ice coating. Have been writing to the owner and making out copies of the regulations now in force.

Tuesday June 14
Blowing strong but moderating towards night with the wind backing to north. All ropes and rigging heavily coated with ice. Weather clearing. Went over for a few minutes to the steamer as we expect to start back to the boats in the morning.

Wednesday June 15
Left the schooner at 8 AM and came down to the floe to the boats. Commander Low and Doctor Borden also came down. They are going to Southampton using their natives and our boats. We got started by noon and came up to about twelve miles below Whale Point. There are six boats' crews of us. Sky overcast, wind moderate from NE. Can go no farther on account of the ice.

Thursday June 16
We got started at 6 AM and worked our way through the ice. Got within four miles of Whale Point but the ice being jammed in we could go no farther, then hauled out. Wind moderate from north, mostly cloudy. Six seal taken.

Friday June 17
Very pleasant and too warm for comfort. We started at 7 AM and working when the tide favored us and slacked up the ice we reached Whale Point

at 10 AM. There is quite a little water on the ebb tide. Light air from NW. Latitude of Whale Point, as determined by Commander Low, 64°12'32"N.

Saturday June 18
Moderate breeze from south to SE crowding the ice in. No water to be seen offshore.

Sunday June 19
Fog during the morning, later becoming clear. We made an attempt to start across but the ice closed in tight so we had to return, then made another attempt at 6 PM when the tide slacked. Got part way across when the ice became tight and we hauled out on a cake of ice. Light wind from NW. Shot a bear.

Monday June 20
Made a start at 6 AM and reached the shore ice which was off from the land about seven miles. Wind generally moderate from NW. In coming across we have had much ice.

Tuesday June 21
Started again at 8 AM and worked along the edge of the shore ice to the SW. We found we were much farther to the north and that the Southampton shore after leaving Cape Kendall the land makes to the ENE about twenty miles (on the north side) then it starts directly to the NNW and at Wager River a person can make out the land on the opposite shore. We got to the south of Cape Kendall but could not haul out on the shore ice as the loose ice was packed in. Hauled out on a cake. Wind moderate NW. Got two ground seal.

Wednesday June 22
This morning we left the cake of ice and worked in to the shore ice, Commander Low thinking they would stop here and look over this part of the country. He gave us some of his provision. We bid them good-bye and with three cheers we started in to look for whales.[7] In about two

7 Low, Borden, and their eight Eskimo sailors spent the next week cruising along the west coast of Southampton Island in the region of Cape Kendall, making short excursions inland to observe rocks, vegetation, animal life, and ancient Eskimo houses. They collected fossils, artifacts, bird skins and eggs (Low, *The Cruise of the Neptune, 1903–4*, 33, 34). On 23 June they constructed a cairn of limestone blocks around a tin box containing this message:

hours after leaving them we saw a whale and though we worked two hours to get him we could not, the ice being thick and the water shallow. This whale was going to the north. The reef which makes to the southwest on the southern part of Cape Kendall extends off about eight miles on which the ice was grounded, and as we had to go outside it made our distance much longer. After getting well around it and up in sight of land again we hauled out at 10 PM. Wind north moderate, morning misty, later partly cloudy.

Thursday June 23
Pleasant though some fog this afternoon. We reached the land about eight miles south of Manico Point. Had a moderate breeze from NE but when within fifteen miles of the land we had to pull the rest of the way.

Friday June 24
This morning before the tide got high enough I went to some old igloo native houses which had been made of stone and the head bones of whales. It is the same place where I first saw the Southampton natives in 1896 and got some whalebone from them. I gathered up some old implements. All the natives are now dead on this island, this being caused by the Scotch station here which employed some 150 natives from elsewhere, who with guns soon made game so scarce that the people of Southampton could not get a living.[8] The weather being stormy we stopped after coming about twelve miles. Rain and fog.

In the name of our Gracious Sovereign King Edward VII and on behalf of the Government of Canada, I renew and take Possession of this Island of Southampton for the use and property of the Dominion of Canada. God Save the King.
[signed] A.P. Low
They then hoisted the flag above the cairn, recited the proclamation, and gave three cheers (A.P. Low, Geological notebook no 2488 [15 June – 1 July 1904], 7. Ottawa: Geological Survey of Canada).

8 The Sadlermiut, who comprised fewer than seventy individuals in the late nineteenth century, were almost totally successful in avoiding contact with other Eskimo groups and with explorers and whalemen, but in 1902 all but five died, as Comer records in his journal entry 9 August 1903. Low and Comer believed that well-armed Eskimos from other regions, relocated to Southampton Island to work for the whaling station, had probably destroyed the subsistence base of the island's original inhabitants and brought about their starvation. Mathiassen, on the other hand, heard oral testimony that they had died from disease carried by the whaler *Active* (see chap 1, n 11). The latter is supported by details in whaling logbooks.

Saturday June 25 Scotch station
Still foggy with light sw wind and calms. Started at 11 AM and pulled down to the station. Of course we did not expect to find anyone here as the station was abandoned last year and the people have gone to Repulse Bay. Got three gulls' eggs (poor), one seal taken. Purple flowers are now making their first appearance. Shovelled away the snow so we could get in the house and then shovelled out the snow from the rooms.

Sunday June 26 Scotch station
Thick fog during the day. At times it would light up for a few minutes. At night a little better but overcast. Moderate breeze from north. There is a framework of small spars erected about forty feet high from which we can keep a lookout.

Monday June 27
At first thick fog with increasing wind from north. Quite cold and damp, rough sea. Hauled the boats up higher. Took some provision from the house.[9]

Tuesday June 28
At first stormy with strong winds from north. After noon moderate and becoming pleasant. I took a walk down towards the point to find out, if I could, how much farther south the island extended. Should think about ten miles south of the house would be the southern point, which would make the latitude nearly 63°00′N. Got seven king eider duck eggs.

Wednesday June 29
Started off at 4 AM to cruise in the boats. Have had light variable winds and calms, very little ice. The prospects look poor as we had ought to see whales if there is to be any. Got one bear and an oujoug (ground seal). Mr Ellis and Reynolds are about seven miles north of us.

Thursday June 30
We are one year out today. We have been off cruising but nothing to be seen. Got about 250 lbs of bread from the house. Moderate breeze from north.

---

9 Captain Alexander Murray of the *Active* had given Comer permission to use provisions left at the abandoned Scottish station (see journal entry 29 August 1903).

Friday July 1
Pleasant. We have been off cruising, have had light winds from WNW. Got one oujoug.

Saturday July 2 Eight miles north of the station
Very pleasant and calm most of the day. What wind we have had has been WNW. It is at this place where there are so many whale-heads, something over forty in a small space and plenty of others near here. Got a few eggs and one bird nest. Mr Ellis is about four miles north of us.

Sunday July 3
Calm all day. We have been off laying around and drifting with the ice. Quite warm.

Monday July 4
We started, two boats of us (Ben's boat and my boat), to go to the north of Manico Point and try for deer but when we got up north we found the shore ice extended so far from the land that the land could hardly be seen, so returned to the south of the point where we could make a landing. Got a number of implements from native graves. Put a cartridge at each grave in payment for goods taken. [10] Near each grave was placed stone which formed a seat for the mourner to sit in when visiting the grave and talking to the spirit of the dead person. The foot of the grave was generally towards the sea.

10 Although Comer (working on behalf of museums in the United States, Germany, and Canada) collected many articles from Eskimo graves during his voyages, including skeletal material, he was always respectful of both the living and the dead. As he explained to anthropologist Franz Boas after obtaining four skulls on Southampton Island during his previous voyage: 'the natives tell me I can take them if I will place a small present in the grave for the dead person – then it will be alright' (Comer to Boas, 4 August 1901, Correspondence relating to Comer acquisitions, 1902–78, American Museum of Natural History, New York). Sometimes Comer placed gifts in the graves of Eskimos he had known well, without removing bones or possessions (see journal entry 26 August 1903).
There is no evidence, however, that such an ethical and considerate procedure was followed by either Low or Borden, who obtained some artifacts and human bones from Southampton Island in June 1904. Their collections, moreover, were not destined for science alone. Borden took a Sadlermiut skull which he possessed all his life (Borden, 'Memoirs,' 92) and while at Fullerton secretly removed an entire skeleton from a grave late at night without consulting the residents of the community, although with the approval of the commander of the expedition (ibid. 85).

Thursday July 5
We have remained here today and have the natives go hunt for deer. They only got one. My men got quite a few eggs, such as king eider, three swan's eggs in one nest (yesterday got two in one nest), also small bird nest with eggs.

Wednesday July 6
Pleasant. We came back to the station. Cruising offshore got one bear and two oujougs. Found Mr Ellis had taken two bears. A little more ice in sight. Light wind west.

Thursday July 7
Overcast with fresh breeze south. The natives saw a large whale which was going south quite fast. To the south, there is much more ice which has come up from the south, probably through Fisher Strait.[11]

Friday July 8
Thick fog with rain squalls. Have not been off. While we have been here I have been able to collect quite a lot of egg shells and native implements. The men have done much to help me in collecting.

Saturday July 9
Thick fog till 11 AM when we went off and cruised till 8 PM. There is considerable ice now. The wind hauled from SE to NW. Raining now and rained much last night.

Sunday July 10
Raining during the night and fog and rain during the day. Had to haul our boats up higher. Did not go off. Wind fresh from NW.. There is but little use of our remaining here.

Monday July 11
Have been off cruising. Quite a little ice – some very large pieces. Got one bear and an oujoug. I had a pleasant time while waiting for the bear

11 A map of Southampton Island drawn by Comer and published in 1910 by the American Geographical Society shows a current flowing northward past Cape Low towards Cape Kendall. North of Cape Kendall it turns to the west across Roes Welcome Sound, then it heads southward along the mainland coast of Hudson Bay (George Comer, 'A Geographical Description of Southampton Island and Notes Upon the Eskimo,' facing p 84).

to come up to me as it evidently mistook me for a seal or walrus as I lay on the ice (it was a large piece of ice) with numerous holes in it. After the bear saw me he went down one of these holes and swimming to one nearer would come up to breathe, then down under the ice and up again, each time only showing the point of his head. In this way he approached quite near and when in the nearest hole to me I arose up and shot it, but did not kill it, when down it went and started to retreat, coming up in the next hole. I shot again and killed it, a large bear. Wind light, variable, mostly east. Got one oujoug.

Tuesday July 12
Stormy during the night and this forenoon. At noon the weather improved and we started to return to the vessel to come by the way of Whale Point. Wind fresh from NE. Had a strong head tide and the weather becoming stormy we stopped about six miles north of station. At night weather improving but we cannot get the boats off till the tide rises.

Wednesday July 13
We started at 2 AM and, by sailing and pulling and with sunshine and rain and calms and strong winds, we got to the land about five miles north of the long reef that makes out from the SE of Cape Kendall at 2:30 AM Thursday morning. I was quite tired and have some cold. We got the boats damaged some in coming in over the shallow places.

Thursday July 14
We remained here today to see if we could get some deer meat but the natives could not see any deer. At low tide the flats make off about two miles – mud and small stone. Two deer and a bear were seen.

Friday July 15
We started at 3 AM, had light airs then the wind sprang up from NE and we stood across. At noon the wind let go and then later sprang up from SW. Got one bear. Arrived at Whale Point at 7:30 PM. Found that our head native Harry had taken a small whale. This he had taken over on the shores of Southampton. He and Sam are not here. The natives are all camped here.

Saturday July 16
Started from Whale Point for the vessel at 5 AM. Had but little wind but a fair tide. Arrived at the schooner at 5 PM. Found the steamer here unable

to get out on account of the ice, which is still solid around us though we came up to the reefs outside the harbor with the boats. Thick fog at times this afternoon.

Sunday July 17
Very pleasant. Lowered all the sails and had them dry, and put the two anchors down through the ice. Spent the evening on board the steamer. They are to try and break out in the morning.

Monday July 18
The steamer went out this morning and while she was going I had a letter put on board by their throwing a line out on the ice and had the letter made fast to it which they hauled on board (the letter for my wife). We gave them three cheers and they returned it. After breakfast we got pro-vision ready and fitted out our boats to return to look along the shores of Southampton from Cape Kendall to the north as far as Point Harding. Arrived at Whale Point at 6 PM. Our sick man had been discharged and took passage in the steamer.

Tuesday July 19
There has been too much wind from NE for us to make a start to cross over. Harry and Sam's boat returned last night from up Wager River way. They had three walrus and some seal.

Wednesday July 20
Blowing a strong gale from north. We can do nothing but wait.

Thursday July 21
Strong winds from north. The Scotch steamer [*Active*] arrived here and we got our letters from home. I had fourteen. The owners sent out ten bushels of onions and ten bushels of potatoes and a force pump which I had sent for.

Friday July 22
Started from Whale Point at 9:15 AM. Had light variable winds. Arrived north of Cape Kendall at 10 PM at high tide and hauled out.

Saturday July 23
Left the beach at 9 AM and following the shore to the ENE about twenty miles where the land was very low, which I think connects with low land on the south shore of the Cape. It is so low that it can't be seen from a

boat either side. The land here made a turn to the NNW, which we followed up about fifteen miles. Hauled out at 10 PM but the tide did not stop rising till twelve midnight. Light wind west. There is a little ice up here.

Sunday July 24
We got started at 9 AM and were more than glad to get away from the mosquitoes. The natives got one deer here. We came up to what I called Point Harding, the water being quite deep near inshore. As we could not haul out till after midnight, we came across over to the west shore below Wager River about twenty-five miles and hauled out at 4 AM on an island. Light winds NE. While drifting near Point Harding saw what I called a good-sized cod fish near the bottom.

Monday July 25
We pushed off again at 11 AM and had winds varying in force with quite a little swell. We reached Whale Point at midnight and the tide being low we had quite a task hauling our boats out up the rocks.

Tuesday July 26
On account of strong winds we did not push off. Wind NE. Salmon are now quite plentiful and are caught by the natives setting nets off from the point of land, the nets being about sixty to seventy feet long.[12] All the native men come down.

Wednesday July 27
Weather improving. Started at 8 AM and came to the schooner. At first had NE winds, then calm, then SE. Arrived at 7 PM, then I went to the custom house and got a permit to shift our anchorage to the outer harbor. We hove up the anchors and found them badly fouled and towed outside the Provision Island. All the native men came down from Whale Point with us – six boats of us. The vessel has been painted while we were gone as I left one man to help the steward.

Thursday July 28
Fresh breeze, at first from NE. We have been getting the vessel in order and getting off bread and flour. The native men being here makes a large gang so that we can do much work.

---

12 Boas's publications, based in part on Comer's information, do not report fish nets
   used in the traditional culture of Hudson Bay Eskimos. The nets referred to here may
   have been introduced from whaling vessels.

Friday July 29
Wind north to NE. We have been getting off provision and water and stowing them away, also painted the boats (three of them).

Saturday July 30
Wind fresh with rain part of the time, wind ENE to NE. Finished getting off the stores and a cask of water which will make us about seventy barrels of water. We are about ready to leave here for a more profitable location, though I do not know just where we will winter but hope somewhere north of here.

Sunday July 31
Fresh breeze from NE, partly cloudy. Got off all the boats, nine of them (there is also one at Whale Point). Took the three ship's boats up and the natives started off for Whale Point with the other six. At night weather clearing. The men have been ashore washing their clothes (or the dirt off their feet).

Monday August 1
We got underway at 2 AM with the wind moderate from NNW, quite clear. Later the wind died out and has been light and variable so that we are hardly up to Whale Point, though we can see the tents and house on the point. Could see our natives sailing up along the coast this forenoon. Partly cloudy.

Tuesday August 2
Light airs and calms during the night and today we took some of the heavy goods of the natives who are to go up to Repulse Bay in their boats. Then they started, six boats of them. Harry and Ben and their families are going up in the schooner with us. We have quite a deck load. This afternoon have a light air from SW, a few rain squalls. Eight PM we are nearly half-way from Whale Point to Wager River. The small head of bone was brought off, which we cleaned today. It weighed 230 pounds.

Wednesday August 3
Light airs and calms. Have come but little ways. We are now a little south of Wager River, have met some bay ice. This afternoon we got four oujougs (ground seal). Could have got more had we cared to spend the time. Light airs SW to north, a little fog this afternoon. Lat at noon 64°45′N.

Thursday August 4
Light airs, variable, and calms. Some fog and quite a patch of ice, which has probably come out of Wager River. We are now a very little north of Wager River. Lat at noon 65°18′N.

Friday August 5
Have had light airs and calms till towards night when the wind increased to a fresh breeze from north. Took in the light sails. Lat at noon 65°35′N. The vessel's leak is 2,500 strokes a day.

Saturday August 6
Fresh breeze during the night from north and the same today, toward night backing to NW, improving. We came up to Repulse Bay and taking the west passage into the anchorage we got jammed in the ice and in the hauling through we got aground at midnight at nearly low water.[13] The steamer [*Active*] was able to get in here this morning. We are still aground 1 AM. Both the ketch *[Ernest William]* and steamer are here. The ketch has one whale taken last fall.

Sunday August 7
The schooner lay aground from midnight till 5 AM, when she floated. We worked cutting through ice to get into the harbor till noon and got through, though we have been in much danger of losing the vessel. We received no help from either of the Scotch vessels. Their boats have been cruising around nearby. How very much different it would have been had they been American vessels and men, in which case they would have rendered all the assistance possible. Moderate breeze west to SW at night.

Monday August 8
Pleasant. Three boats have been cruising and I have remained on board with my boat's crew getting ready to have the boats start for Lyon Inlet tomorrow. The six boats arrived this evening, having sailed up from Whale Point. Moderate breeze south. Sergeant Dee of the police force arrived with a boat from Cape Fullerton. Mr Cleveland is now working for the ketch (Captain John Murray) and is now at Lyon Inlet.

13 The anchorage in Repulse Bay was among the dozen-odd islands of the Harbour Islands group. The main approach was through the southern entrance but in some conditions of wind, sea, or ice captains chose the eastern or western gaps, which were less than 400 yards wide.

Tuesday August 9 North shore of Frozen Strait
We left the schooner at 7:30 [AM], six boats of us, four of them to try and get to Lyon Inlet while I with another boat intend to go to the Duke of York Bay. We are now hauled out on the SE shore of Repulse Bay. Moderate winds south. This morning I put the chronometer on board the steamer to be taken to Scotland and be repaired and be brought out next year. After reaching this place I made an attempt to cross over to the other side of Frozen Strait but the wind coming in ahead and meeting some ice we returned to the other boats. No heavy ice.

Wednesday August 10
The weather has been bad with rain and fog, at night improving. Wind moderate from SE.

Thursday August 11
We got started at 1 AM, four boats going towards Lyon Inlet while I with another boat (native) came across to the island on the north of Southampton. This island is thought to be a part of Southampton but there is a narrow strip of water even at low tide which separates it.[14] We met some of the natives here who were carried over there on the ice in the spring of 1902. Three deaths have occurred among them and they lack many things.

Friday August 12
We made an attempt to get through between Southampton and the large island and though we got nearly into the Duke of York Bay we were prevented by the ice forming a jam. The whole of the bay was covered full of ice. It would hardly be safe for a vessel to attempt to go through this passage though it might be done. We landed on Southampton and got three deer. We returned to where we met the natives. They wish us to take them back with us to the mainland. Light airs and calms.

Saturday August 13
This morning we took the natives in our boats – fourteen of them, two dogs and all their belongings – and brought them over to the mainland. In coming across, we struck two narwhals but lost them as the harpoon drew out. Got one ground seal. When we got into the current of Hurd

14 The narrow channel between Southampton and White islands was labelled 'Boat Channel' on Comer's 1910 map (ibid.). It has subsequently been named 'Comer Strait.'

Channel, the ice running very swift, a kayak which was being towed by the other boat got caught in the ice and lost. Hauled out at 11 PM.

Sunday August 14
Pushed off at 6 AM and went around to Gore Bay to a point of land marked on the chart as Farhill [Point]. The native name is Igloo-ju-ack-talic. We found our other boats there. Being unable to get around to Lyon Inlet we brought the people we had with us and landed them here. Our prospect of making a good voyage grows smaller as time goes by and we see nothing. At present I do not see any place around here where we can winter. The height of this point is 225 feet. The second hill a little farther back is 350 feet high and the beaches extend to the top, showing that at one time the land was under water.[15]

Monday August 15 Gore Bay
Have not pushed off today as there is much ice and we can do but little. Two boats of natives have gone to the head of the bay to hunt deer on an island not laid down on the chart. Wind moderate SW. It has moved the ice a little.

Tuesday August 16
No great change in the ice. Have told the three native boats to go and hunt deer and then go to the schooner and bring provision to us. We took the natives we brought from Southampton and landed them near [the] salmon streams. Wind south.

Wednesday August 17
Wind SE fresh. The ice is closing in. Some rain and fog. There are the

---

15 Following the retreat of the thick continental ice sheet from northeastern North America several thousand years ago, much of the land, having been depressed by the weight of ice, was below sea level and was therefore invaded by the sea. In the subsequent rebound of the land – a process which continues even today – evidence of post-glacial marine transgression (sea shells, marine mammal skeletons, raised beaches, and so on) has been found at impressive altitudes. In the region of Lyon Inlet and Repulse Bay the marine limit is approximately 200 metres higher than the present sea level (John T. Andrews, *A Geomorphological Study of Post-Glacial Uplift with Particular Reference to Arctic Canada* [London: Institute of British Geographers 1970], 62).

remains of ten old sod and bone houses here but the natives have no remembrance or tradition of people having lived here.[16]

Thursday August 18
Wind fresh from east, cold and raw. Can do nothing.

Friday August 19
No change in the weather, the ice being close packed.

Saturday August 20
Light variable airs easterly.

Sunday August 21
Came over to the west shore through the ice. It took all day, all the men getting quite wet. Wind light SE. Some snow and fog.

Monday August 22
Fog, rain and snow. Wind SE moderate. We cannot move to any advantage and cannot get our clothes dry.

Tuesday August 23
Blowing strong with some rain and snow. Supper of hardtack and coffee. Height of hills at this place 650 feet.

Wednesday August 24
Very strong winds from north during the night. Today moderate. Light rain most of the time. The ice is making out.

Thursday August 25
The wind coming to the SE we thought it best to get out to where we could reach the vessel should we want, so we worked the boats through the ice around to Hurd Channel, it taking all day.

Friday August 26
At 5 PM we started to return to the schooner and at midnight hauled out

16 These were probably houses of the preceding Thule Eskimo culture which had dominated the Canadian Arctic after about 1000 AD, during a period of warmer climate and greater availability of sea mammals. The circular houses were made of boulders, sod, and the bones of whales (see Robert McGhee, *Canadian Arctic Prehistory* [Ottawa: National Museum of Man 1978], 83–102).

at the SE cape to Repulse Bay. Wind east fresh. Fog and mist during the day.

Saturday August 27
We arrived at the vessel at 6 PM. Found her alright though they have been dragged around the harbor by the ice coming in. Our three boats left here Thursday so that we must have passed them last night. The steamer took one whale down in Frozen Strait while we have been gone. Wind east to NE.

Sunday August 28
Moderate weather, light winds from NE to north. The men have had today to wash their clothes. Went on board the smack and the steamer this evening.

Monday August 29
Blowing strong from the north. As we were laying too near the west side of the harbor we ran a line to the steamer and then took up our anchor and dropped astern of the steamer. At night a little more moderate. Had the graphophone out this evening. Got out provision.

Tuesday August 30 Hurd Channel
We left the schooner at 8 AM and started for Lyon Inlet. We met our three native boats returning, having been passed by us during the night when we were coming to the vessel. Took them back with us but found the ice so close that we had to haul out after getting into Hurd Channel. Have had fresh breeze from north.

Wednesday August 31 Hurd Channel near Gore Bay
This morning, the ice still being close, I told the native boats to work back to the vessel while we three would wait for the ice to slack up, which it did later in the day, and we came down to near Gore Bay, which is full. We are now waiting to get over to Vansittart Island. Wind north fresh. Got one seal.

Thursday September 1
We got through the ice and got to the SW side of the island (Vansittart) but cannot go but little ways as Frozen Strait is full of ice. This season there seems to be a lack of deer on this end of the island. Our prospects of making a voyage grow less each day. We are now fourteen months out

and I have only seen one whale.[17] We have but about sixteen days more to work this season.

Friday September 2 Vansittart Island
We pushed off at 4 AM and made an attempt to work to the southward but could only advance about four miles. As far as can be seen from the highest land there is no water to be seen. We then worked back to last night's camp. Light airs, variable. Got two mice – these I get for the American Museum.[18]

Saturday September 3 Southeast cape of Repulse Bay
We started at 8 AM to return to the schooner as the wind is from the NE strong. Came up in Frozen Strait through much ice. Hauled out at 6 PM. Wind east.

Sunday September 4 Blue Lands salmon stream
We pushed off at 5 AM with light easterly winds. After passing the island near the SE cape I raised a small whale and after working around it for some four hours with light airs and calms I got a chance to go on and, though it was a splendid chance to have killed the whale, my boatsteerer missed it by throwing both irons down [the] side of it and the whale soon after went into the pack ice, apparently not gallied.

Monday September 5 On board
We started at 7 AM and cruised back over the ground, as far as the ice would let us, where we saw the whale yesterday but the wind being SE

17 Comer's statement indicates the extent to which the Hudson Bay whale stock had been depleted during half a century of whaling. In 1860–1 the *Syren Queen* had sighted whales on forty days. Her ship's logbook records sightings of thirty-six individuals, and on an additional twenty-nine occasions the numbers sighted were described as 'some,' 'several,' 'lots,' or 'plenty.' The total number sighted probably exceeded 200, of which twenty-one were killed (Manuscript logbook of the *Syren Queen*, 1860–1, Kendall Whaling Museum, Sharon, Mass).

18 Just before the *Era*'s departure for Hudson Bay Comer had received a list of items which the Department of Mammalogy and Ornithology at the American Museum of Natural History wished him to collect. It included 'Lemmings and Mice – three species, about 10 of each, at 50 cents each' and '20 Ground Squirrels at $1.00 each' (J.A. Allen to Comer, 18 June 1903, George Comer Papers, East Haddam, Conn). These were probably the brown lemming (*Lemmus trimucronatus*), Greenland collared lemming (*Dicrostonyx groenlandicus*), tundra redback vole (*Clethrionomys rutilis*), and arctic ground squirrel (*Citellus parryi*).

had moved the pack up the bay. Then we worked towards the vessel. Brought over four natives of the Scotchman who had walked from Lyon Inlet. They had been two days in coming. They say they have seen nine whales in the ice. Quite a little snow fell last night. This evening a large whale was seen near the harbor by our natives.

Tuesday September 6
Have been off cruising with our six boats. The Scotch vessels have about twelve so that it makes out quite a fleet. Wind variable but mostly NE.

Wednesday September 7
Moderate weather, breeze from north. Went off cruising this forenoon but I came back with my boat and have been at work on board the vessel. Mr Ellis and Reynolds have gone over to the Blue Lands to stay one night, also Sam's boat. Captain Murray of the steamer is quite sick – stomach trouble.

Thursday September 8
Have been at work on the vessel with my boat's crew overhauling the furs. Put our whalebone on board the steamer – 37 bundles. Its weight was 1,853 pounds. All the boats are here tonight. Fresh breeze N to NW.

Friday September 9
The five boats have been off cruising while I have been on board with my boat's crew working. We got off from our house ashore (which we built last voyage) about a ton of coal which we left there. Fresh breeze from northwest to north.

Saturday September 10
Fresh breeze from north. The boats have been off cruising but nothing seen. I shall now let the natives go off to hunt deer and think it best for us to go down the Welcome and winter at or near Cape Fullerton as I cannot see any better way to conduct the voyage. Wrote a few letters and put them on the Scotch steamer. Got off a cask of water.

Sunday September 11
The wind is increasing to a fresh gale. We have now got everything ready to start and go down the Welcome but I wish to take a more moderate time so we can stop and take the furs that are at the Wager River station, which have been collected by Mr Cleveland for our owners, as he is now

employed by the Scotch station here and is whaling for them. Paid our natives in ammunition and four boats are to remain here and continue to look for whales till the season closes. Wind north with snow. Got off two casks of water. We are to take six boats with us. Shall send one of my men home by the steamer as he is not well enough to live out the voyage, I fear.

Monday September 12
Still blowing strong from the north, a few snow squalls. Took the man to the steamer who is to go home in her. I told Captain Murray that in case Mr Cleveland did come before he left and should go home with him, not to stop at the Wager River house to take the skins as I would attend to that in the interest of Mr Luce.

Tuesday September 13
Weather improving. We started off at 10 AM, taking with us two families of natives, Ben and Jimmy's. Had the wind NW, some swell to start with but later smooth. Some ice on the Southampton shore. The steamer came out soon after us but at night she is hull down astern of us. Wind light. We are working down the Welcome and think we will have to winter again at Cape Fullerton.

Wednesday September 14
Light moderate weather during [the] night and the same today. We stopped at the station which Mr Luce had at Wager River and took what skins were there – 350 musk-ox, 16 bear, 14 wolf, 6 wolverine, also 1 vice, 1 grindstone, 6 lantern globes, 1 shoulder gun and 2 dozen bars soap. The steamer passed us at 6 PM. Wind south light. Two of our men are sick with cramp in their stomach. Every one of the skins is spoiled by being wet. Many of them the maggots are thick on them, not good skins in the first place. Evidently all the best skins have been taken out, including all the fox skins.

Thursday September 15
Light airs and calms during the night. Today have had variable light winds from west and north, at night increasing from the NE. Snow squalls this afternoon. At 8 PM shortened sail and came to near Yellow Bluff. Not bad weather but wish to work along tomorrow if the weather is good.

Friday September 16
Last night we lay to near Yellow Bluff. Moderate breezes with snow

squalls from north to NE. Today we came along, keeping a lookout for whales, and came down to Cape Fullerton, where we arrived at 1:30 PM. Have had strong breeze from NE. Found the Canadian steamer *Neptune* here, having arrived this morning. Major Moodie left her at Port Burwell and returned to Canada in the supply steamer. We are in the outside harbor. I spent the evening on the *Neptune*. The sick man who was taken home for us by the steamer was given about $30.00 by the steamer's crew when he was transferred to the *Erik*.[19]

Saturday September 17
Light moderate weather, wind hauling from NE to SE, then to south, and snowing at night. Came inside to the inner harbor and entered at custom house. Our men got five rabbits and one duck, and our natives which we brought down from Repulse Bay have taken a boat and have gone up towards Whale Point. Took supper on the steamer.

Sunday September 18
Moderate SW winds with some snow last night. Winds the same today. At 6 PM calm. At 9 PM wind suddenly came out of the NE fresh, with snow. Let go the second anchor. Took supper and spent the evening on the steamer. Commander Low gave me some newspapers, also twenty small books printed in the Eskimo language for my natives.[20]

Monday September 19
Snow squalls, wind fresh from NW. Have landed some spare lumber and spars and also boats and topgallant yard and squaresail yard. Getting ready for winter. Put fifteen coils of tow line in two casks ashore.

19 In July the *Neptune* had sailed out of Hudson Bay to cruise northward among the Arctic Islands. On her way she stopped at Port Burwell and picked up provisions, coal, and mail from the supply ship *Erik*, which then returned to Newfoundland (Low, *The Cruise of the Neptune, 1903–4*, 40, 41).

20 A system of syllabic writing had been devised by the Reverend James Evans in 1840 to facilitate missionary teaching among the Cree Indians. The Reverend E.J. Peck subsequently introduced the system to the Eskimos of Baffin Island following the establishment of a mission at Blacklead Island, Cumberland Sound, in 1894. The books referred to by Comer were probably prayer or hymn books printed by the Church Missionary Society in London, and obtained by the *Neptune* at Cumberland Sound (Fabien Vanasse, 'Relation Sommaire du Voyage de l'*Arctic* à la Baie d'Hudson, 1904–5,' 32).

Tuesday September 20
Wind NW, varying in force, with snow squalls. Three partridges were taken. Landed some small things. Took supper on the steamer. Commander Low gave us twenty-four books from his ship's library and other things such as nails, screws, putty, and little things such as they could spare.

Wednesday September 21
Snow squalls but generally good weather, wind varying from NW around to west, then SE and becoming overcast. They are having a dance at the house ashore. Five rabbits and eight ducks taken today, Doctor Borden going with me and shooting the ducks.

Thursday September 22
Stormy day, wind fresh from ENE to NE. At night clearing, wind north moderate. Two saddles of deer meat brought in by a native known as Smiley. Took the graphophone down in the forecastle and later spent the remainder of the evening on board the steamer.

Friday September 23
Moderate cloudy weather, wind west NW. Four ducks were taken today. The steamer is preparing to start tomorrow. I have been writing to friends at home. Spent the evening on board the steamer and took supper.

Saturday September 24
A stormy day with the wind SE, thick snow most of the time. At night the wind shifted to NW and the weather cleared up, wind light. I spent the evening on the steamer. Have been writing letters to friends at home.

Sunday September 25
Wind SE fresh with snow squalls during the forenoon, afternoon weather improving. The steamer started at 1 PM on her return voyage to Halifax. I put my letters on board. We gave them three cheers as they passed which they returned with a will. Three policemen have been left at the barracks, which includes Sergeant Dee.

5

Preparing for Winter
(26 September – 31 December 1904)

Monday September 26
Snow squalls most of the time. We are now building the house over the after part of the vessel. Took up the small anchor, and unbent the mainsail and tied it up with the foresail. Wind SE fresh.

Tuesday September 27
A stormy day, wind SE with snow and mist. Have not been able to do any work towards building the house over the deck. Let go the second anchor – a little swell gets in here from outside.

Wednesday September 28
Dull, misty, and light rain continually. Wind fresh from SE to SSE at night. Took up the small anchor. Have been working on the house. Some of the Kenepetu natives have been on board.

Thursday September 29
Still the wind remains SE and rain most of the time. Are unable to go ahead with the house. A number of the natives came off to hear the graphophone this evening. Raining steadily.

Friday September 30
Blowing very strong from ENE with a blinding snow. Can do nothing but attend to the anchor chains. We have but little room to swing astern of us as we are near the rocks, but the water is good depth. Such a storm must be hard upon the natives living in tents, which are far from being waterproof. Have dealt out clothing to the men for winter use. Fifteen months out today.

Saturday October 1

Weather clearer and wind more moderate, but still strong from NE. Have been able to work on the house. Got pretty well along with the boarding up. The vessel leaks a little more now, probably on account of its being rough and the leak or part of it being above the water line.

Sunday October 2

Weather more moderate. Some of the Kenepetu natives have been off to the vessel. The vessel's leak today was 1,200 strokes, which is more than it has been before this last storm. Winds still NE.

Monday October 3

Light winds from north. Young ice is making and drifting out. Six ducks taken. The Kenepetu natives left here today to go deer hunting. When we were at Repulse Bay five musk-ox skins were brought to the vessel. I gave the natives what they wished for them in ammunition and have today turned them over to Sergeant Dee of the police force here. At 9 PM the temp is 12°. Seven natives are retained by the police – this includes only two men.

Tuesday October 4

Pleasant with moderate breeze from NW. We have the house now closed in and the windows all in – eight of them, five of them being at the stern. Our people were off today gunning but had no success. Temp through the day 12°, at 9 PM 8°. Young ice making.

Wednesday October 5

Pleasant with moderate breeze from NW. Took up the small anchor. The vessel is held in the young ice though the harbor is not frozen over. Six partridges were taken. Have been caulking the seams of the house.

Thursday October 6

Cloudy but light weather. Built on the meat locker a place about three feet square and six feet high where we put fresh meat when we have extra. Four ducks taken.

Friday October 7

The wind hauled to the SE during the night, which broke up the ice which had formed nearly all over the harbor. Let go the second anchor. Winds moderate, some snow during the night.

Saturday October 8
Cloudy with moderate breeze hauling from south to west. Hove up the anchor and dropped back to have a little more room. This evening I took the phonograph down into the forecastle and gave a number of pieces for the men's benefit. Saturday is our day for getting out provision.

Sunday October 9
Pleasant weather, wind NW moderate. Young ice making and drifting out. Sent a boat in and had a piece of the pond ice brought off, which is now ten inches thick. Have got ready to go in tomorrow and cut our winter supply. Temp 5°.

Monday October 10
Moderate breeze from NW. Today we had our winter ice cut. This will last us till we can get water in the spring. We have waited most too long before cutting as the ice is now about twelve inches thick. Opened the slop cask and took out clothing for winter use. Young ice covering the harbor. Temp during day 5°.

Tuesday October 11
The wind backed to the south and the weather has been much milder. The harbor is completely covered over with young ice. Wind fresh. At night, 10 PM, is hauling back to west though the thermometer is and has been 32° the latter part of the day.

Wednesday October 12
Thick fog last night and today. At night the wind sprang up from NE and has brought the ice that has lain to the NE of us down around us, so we are now fast in the ice, which is quite strong. The weather has been warm today but is now becoming colder, dropping from 32° down to 12°. At 10 PM we took up the small anchor. As I am getting to be quite bald I have commenced having my head shaved, hoping it will induce the hair to grow.

Thursday October 13
Pleasant with a moderate breeze from NW. We sawed the ice up to the anchor and then placed the vessel in position for the winter heading due north, the ice being about four inches thick. Temperature at night 11°.

Friday October 14
Strong breeze from SW with a much warmer temperature, some hail and

rain. Some swell gets in here at high tide, which causes the ice to move, and it had softened up considerably. Today let go the anchor for fear the ice would break up. At 3 PM a steamer could be seen to the south. She seemed to be trying to find the entrance to the harbor. She is barkentine rigged with smoke stack well aft. The wind is light from the west, sky overcast.

Saturday October 15
This morning, the weather being dull and wind east moderate, a boat came in from the steamer (*Arctic*), which was anchored outside the reefs last night, then went to the police house and later a note was sent to me from Major Moodie, who is on the steamer, asking for natives to come out and help pilot the steamer in.[1] But as there are none here I could not

---

1 The *Arctic* went north to relieve the *Neptune* and continue to support Canadian sovereignty, showing the flag and enforcing the law in Hudson Bay and the Arctic Islands (see appendix K(B). Superintendent Moodie had accompanied the *Neptune* from Fullerton to Port Burwell in July 1904, but while there had suddenly decided not to remain on board the vessel (which was about to cruise through Davis Strait, Baffin Bay, and Lancaster Sound) but rather to travel south on the supply steamer *Erik*, go to Ottawa to discuss various aspects of the northern work, and return to Hudson Bay with the relief ship *Arctic* in the autumn. According to the Toronto *Mail and Empire* of 21 August 1904, Moodie was very critical of the *Neptune* expedition. It had been badly fitted out, he is alleged to have told reporters; the crew had been either unsuited for the job or simply incompetent; there had been personal friction 'from the outset'; Commander Low, a renowned explorer, had carried out no explorations whatever; the only explorations had been by the photographer, Caldwell, but he appeared to be keeping gold discoveries a secret for personal gain; the expedition had 'collapsed miserably'; lawlessness and crime were 'on the increase.' According to the newspaper Moodie would seek an extension of his powers before going north again, and he would try to 'prevent political and other appointments destroying the value of a mission so fraught with great possibilities for Canada.' He intended to establish two more police posts, one in Hudson Bay and one in Cumberland Sound, and he needed a force of thirty policemen, with two ships (RCMP Records, vol 280, file 707). Earlier he had recommended to the comptroller of the RNWMP that his position should have the 'rank and title of Lt. Governor' (Moodie to White, 1 July 1904, p 25, RCMP Records, vol 281, file 716), and that his force should be provided with a patrol ship mounting two rapid-fire guns, both measures calculated to engender a greater amount of respect on the part of the whaling interests (Moodie to White, 27 July 1904, p 2, ibid.).
   Moodie later denied that he had been so critical of the *Neptune* expedition (Moodie to White, 30 August 1904, RCMP Records, vol 280, file 707). Nevertheless, when he returned to Hudson Bay on the *Arctic* in 1904 he did so with more power and authority than before. Aside from being commanding officer of an expanded police detachment at Fullerton, he was commander of the entire expedition – vessel and all – as well as fishery officer, magistrate, and customs officer.

send any and he has in his employ one man and the interpreter. They both went off with the boat. Soon after leaving here the weather became foggy and the steamer has been unable to come in. It is reported that she will winter here. Shall know more about her plans when she comes in. Winds light from east. Temp 32° during the day.

Sunday October 16
The wind was fresh this morning from NE. The steamer came in and anchored near us, the ice being easily broken by the steamer. We set our colours for them. No communication made today. Wind light at night and still NE. Temperature falling to 17°. Major Moodie has his wife with him.[2]

Monday October 17
Wind south moderate, sky overcast. Today Major Moodie requested the use of the carpenter for a few days as they are going to build a larger house ashore for the police force.[3] I have allowed the carpenter to go to work for him. He commenced this noon. The major sent over as a present to us one barrel of apples, one box of oranges, one bunch of bananas, and four bags of new potatoes. These have been touched with frost which was unavoidable. These are things which we are in a position to appreciate. I went over to thank him and met his wife, also Captain Bernier. Walked over on the ice. Temp 29°. One partridge shot.

Tuesday October 18
Wind southeast to NE, overcast. Temp 32° during the day, at night 29°. We enjoy the apples and oranges, in fact all the things Major Moodie

2 In 1903 Moodie had requested the comptroller of the RNWMP that his wife be permitted to join him in the north. If this could be arranged, he wrote, then 'would it be possible for one of my boys to enlist as a "special" for such time as I remain here and come to me if he wanted to do so. He would look after everything better than a Constable ...' (Moodie to White, 9 December 1903, p 6, RCMP Records, vol 281, file 716, part 1). In October 1904 Moodie returned to Hudson Bay on board the *Arctic* accompanied by both wife and son.
3 Because the police detachment in 1904–5 consisted of eleven men, and Moodie's wife and son had to be housed as well, the facilities of 1903–4 (see chap 4, n 4) had to be expanded. The barracks of the first winter was floored with lumber and insulated with asbestos and oiled canvas; it then became the officers' quarters. A new barracks was constructed for the men approximately thirty by fifteen feet in area and insulated in the same manner (J.D. Moodie, 'Report of Superintendent J.D. Moodie on Service in Hudson Bay ... 1904–5,' p 10). The men of the steamer christened Moodie's new residence 'Le Chateau' (Fabien Vanasse, 'Relation sommaire du voyage de l'*Arctic* à la Baie d'Hudson, 1904–5,' p 45).

sent over. It seems strange that we should be having fruit and vegetables now that have been raised this last summer. We have been at work making a small boat to be used at the floe sealing this winter – intend to make two. One rabbit and five partridges taken. We are unable to get ashore without the aid of a boat. The vessel's leak is 300 strokes a day.

Wednesday October 19
Cloudy with light winds from north. Temp during the day 28°. Four salmon, three partridges and one rabbit taken today. Light snow at times. I have been working on the boat that we are building for sealing.

Thursday October 20
Cloudy. Light air northwest hauling to NNE. Temp 22° this afternoon. Finished one boat and have commenced another. Though the ice is strong enough to walk on from one vessel to the other still we cannot get ashore without a boat.

Friday October 21
Pleasant with a light breeze from north. The weather has been colder, temp being 12° during the day. Got the second boat nearly finished. Three of the young men from the steamer were over a short time today. They say that they are to remain out for three winters but have no scientific people on board.

Saturday October 22
Pleasant and growing colder, the thermometer falling to 4°. We took up the anchor and placed the vessel in position for the winter heading due north. One man broke through some new ice and lost his shovel. The outside harbor is now frozen over. Wind fresh from NW. Two of the gentlemen from the steamer were over this evening. Very agreeable. Mr _____ and Mr _____ .

Sunday October 23
Pleasant with fresh breeze from NNW, temp 5°. This evening I spent the evening on the steamer *Arctic* with Captain Bernier. I was quite well pleased with him. He had sent me over during the day a card and pamphlet entitled *The Canadian Polar Expedition* which Captain Bernier is trying to get up for himself to take charge.[4] He also sent me an electric pocket light, a very useful instrument in reading the thermometer after dark.

4 Although Joseph-Elzéar Bernier's seafaring experience on a hundred or more deep-sea

Monday October 24
Pleasant with fresh breeze from NW. Painted the two sharpies with coal tar and pitch melted together and put on hot. Got our ice sled ready. Should have cut our ice for the windows but found it was not clear, so cut a spot clear of ice and will wait for it to freeze to the thickness of five inches, then cut it in cakes large enough to cover outside of each window. By doing this it helps keep out the frost. This of course is fresh water ice. Temp 2°. We could ask for nothing so far as comfort is concerned, if we could have been successful last summer in whaling.

Tuesday October 25
Pleasant with a light breeze from NW. Temp 8 AM − 2°, 4 PM 2°. Thickness of the ice 10½ inches. Put the two whaleboats ashore. Cannot land anything heavy at high tide, the ice being broken.

Wednesday October 26
Pleasant, wind moderate from NW. Temp − 6° at 8 AM, 4 PM − 2°. I took

---

voyages had been in the temperate and tropical regions, he had been fascinated for more than thirty years by the literature of polar exploration and obsessed since 1895 with the idea of leading an expedition to the North Pole. His plan was to replicate Nansen's 1893–6 drift voyage in the *Fram* across the Arctic Basin from Siberia, but following a course about 300 miles farther east. When the ship reached its farthest north a party would attempt to reach the Pole over the sea ice by sled. The 1901 circular on the Canadian polar expedition appealed for financial contributions. It claimed the patronage of the governor-general as well as the authorization of the government of Canada and the Quebec Geographical Society. It emphasized the necessity for Canada to participate in the international effort to reach the Pole, stressed the importance of the scientific results that would arise out of the expedition, and extolled the leadership qualities of Bernier. In the spring of 1904 the Canadian government purchased the German ship *Gauss*, renamed her the *Arctic*, and commenced outfitting the vessel at Quebec City under Bernier's supervision. But at the last moment the government backed down, deciding that its first priority must be the assertion of Canadian authority in Hudson Bay and the Arctic Islands. The North Pole expedition was cancelled and Bernier was ordered to proceed with the *Arctic* to Hudson Bay to relieve the *Neptune*.

On the *Arctic* Bernier was relegated to the position of sailing master, responsible for the ship's navigation and safety but under the command of a landlubber, and a policeman at that – the officious Moodie. To his great credit Bernier accepted the circumstances with equanimity and continued to act faithfully in the service of Canada, contributing in this and in subsequent voyages to the security of Canadian sovereignty in the north. Despite the frustration and disappointment of 1904 Bernier continued to nurture his dream of a polar expedition, as this journal entry of 23 October indicates (see Yolande Dorion-Robitaille, *Captain J.E. Bernier's Contribution to Canadian Sovereignty in the Arctic*, 16–38).

a tramp off on the ice. There is but little we can do now. Opened a cask of beef. We have had little intercourse with the steamer and at present think it is best not to have, as it looks as though the major feels too much above us. He certainly feels as though he was of much importance. While we read much that is in favor of Canada's police force, I can see that they have their share of poor men as well as others. At present we feel it will be better not to be too friendly with them but hope by spring time to improve our opinion of them.

Thursday October 27
Pleasant though becoming slightly overcast. Wind light from north, weather becoming milder. Several of us have been off gunning and as Mr Ellis did not return at dark we feared some mishap had befallen him. We went out and searched for him – a party from the steamer and a party from our vessel. Found him at 10 PM. He was alright. I think he must have lost his way, though he says not but had been chasing a wolf. It is now midnight and a little snow falling. Temp 3°. After finding Mr Ellis he was taken to the steamer where lunch was spread. During the search the men carried lanterns and guns. Major Moodie sent over quite a lot of cabbage, turnips and other vegetables – all frozen but will prove quite an addition to our stock and health.

Friday October 28
Pleasant though partly cloudy. Went out to try and get a seal but could not get near enough. Moderate breeze from NW. Temp 8 AM −4°, 4 PM 5°.

Saturday October 29
Pleasant with light NW wind. Temp 8 AM −4°, 4 PM 4°. Two partridges were taken today. One of the men while out hunting fell and broke the gun stock. Had a dance this evening. Major Moodie and Captain Bernier and many of the crew, as well as a number of the police force, were over. There are not many natives here at present so that there were only about six ladies (squaws). Most of these were quite old but all had a very enjoyable time.

Sunday October 30
Cloudy with south winds light. Temp 8 AM 28°, 4 PM temp 29°. I took dinner on board the *Arctic* with Captain Bernier and spent part of the afternoon with Major and Mrs. Moodie. Was well entertained by them.

Monday October 31
Pleasant with moderate breeze from north. Thickness of ice 14½ inches.
Temp 8 AM 16°, 4 PM 12°. Captain Bernier came over and spent the evening.
Three rabbits, three salmon and one partridge were taken today.

Tuesday November 1
Wind moderate from NE but thick, cold fog. The ice is now fifteen inches
thick around us. Landed eleven tierces of molasses, one cask of coffee,
one cask of powder marked no 1 (blue paint). Got off two sled loads of
ice as our fresh water is now all gone. All well. Temp 8 AM 9°, 4 PM 15°.

Wednesday November 2
Dull weather. Wind south. Temp 8 AM 25°, 4 PM 27°. Put ashore ten casks
of bread, three casks of flour. Went over to the steamer a little while this
evening.

Thursday November 3
Overcast with light winds south to SE. Have been getting provision and
stores arranged in the vessel for winter use. Landed some spare stores.
The vessel's leak is on the increase, today being 560 strokes. One rabbit
taken. Temp 8 AM 27°, 4 PM 28°.

Friday November 4
Overcast with a little snow falling, wind SE light. Temp 8 AM 28°, 4 PM
30°. Two families of our natives came back today. One of the women –
Shoofly – has a heavy cold on her lungs with quite a fever, had to be
helped off the sled and on board the vessel.[5] The doctor from the steamer

---

5 Shoofly was named by whalemen after a song popular at the time (Alfred Copland,
  personal communication, 11 Dec 1982). Descendants give her native name as Niviatsi-
  anaq (Bernadette Driscoll, personal communication, 30 Nov 1982). Fabien Vanasse,
  on board the *Arctic* in 1904–5, gave her name as She-u-shar-kin-neck, and described
  her as an Aivilik woman of thirty-five years, one of the two wives of Ben (Arb-lick),
  aged about fifty (Fabien Vanasse, 'Relation sommaire,' 98). Ben had worked as
  whaleman and hunter for Comer for a decade or so and was friendly enough to allow
  Shoofly to live with Comer on board the *Era* when the ship was in Hudson Bay, but
  the relationship was frowned upon by some of the government personnel. Borden saw
  it as a 'very bad influence on the members of the crew' (Lorris Elijah Borden, 'Memoirs of a Pioneer Doctor,' 66).
     Whether influenced by Comer's example or not, some men of the government
  expedition established intimate relationships with native women at Fullerton Harbour.
  In the autumn of 1904 a woman gave birth to a child sired by a sailor on the *Neptune*,

Arctic has been over twice and is now taking care of her. Three saddles of deer meat and thirty salmon were brought. This party have done very well in hunting, getting something like forty deer in the neighborhood of Whale Point. These natives have taken fifty deer in all.

Saturday November 5
Cloudy. Wind NE moderate, temp 8 AM 26°. Built on the porch to the doorway. Sent over last night to Major Moodie one deer ham and four salmon. So far we have had little snow so that the natives are still unable to build their snow houses, but live in houses made of slabs of ice and covered over on top with a tent.

Sunday November 6
Still cloudy, wind hauling from NE to SE light. The sick woman is slightly better, though still in a dangerous condition. The doctor comes each

---

and men of the Royal North-West Mounted Police detachment at Fullerton are said to have fathered six children between 1903 and 1910 (George Comer, 'Record of Births, Cape Fullerton, Hudson Bay,' Comer Papers, East Haddam, Conn).

Not all the sexual relationships involved consenting adults. In January 1904 Constable Jarvis was charged by Moodie with taking 'a native girl into the [barracks] room for the purpose of prostituting her, she being of the age of 14 years or under' (Moodie, 'Copy of Daily Diary,' 43). Moodie sentenced the man to six months imprisonment with hard labour but remitted half the sentence because he felt the girl, 'altho' young, was by no means innocent. She and virtue had been separated for many a day' (Moodie to White, 1 July 1904, p 14, RCMP Records, vol 281, file 716). Evidently the punishment was not an effective deterrent; a year later Constable Jarvis was on light duty because of gonorrhoea (Moodie to White, 30 June 1905, RCMP Records, vol 302, file 747).

In 1906, when a parliamentary committee was investigating the circumstances related to the purchase and disposal of supplies for the *Arctic* expedition, questions were raised about the conduct of police with Eskimo women at Fullerton Harbour. The prime minister himself questioned the comptroller of the Royal North-West Mounted Police (Laurier to White, 29 June 1906, RCMP Papers, vol 320, file 543). He in turn asked Superintendent Moodie to comment on the allegations of immoral behaviour. Moodie replied that women had been allowed on board the *Arctic* only for dances, lantern shows, and concerts. They had been prohibited from going below and were always 'seen off the ship as soon as the entertainment closed.' Igloos had been placed out of bounds to the men. He acknowledged only that some 'isolated cases' of immorality 'may have occurred' (Moodie to White, 30 June 1906, ibid.). The comptroller then informed the prime minister that during the expedition there had been 'no conspicuous immorality – nothing more than what always occurs between white men and native women, in spite of all precautions' (White to Laurier, n.d., ibid.). Whether or not the prime minister felt reassured by this statement is not recorded.

morning and evening and only allows her to take malted milk. We still have a few oranges left and those taste well to her.

Monday November 7
Still the wind is south with a very little snow and rain. The sick woman is a little better, her temperature being nearly normal. Captain Bernier came over and spent the evening.

Tuesday November 8
Cloudy with a light wind west, moderate weather. At night the wind hauled to the NE and is increasing to a strong breeze. Major Moodie came over today and invited us all to come over to the steamer tomorrow evening. Tomorrow, being the king's birthday, is kept as a holiday by the Canadians. Temp 4 PM 20°.

Wednesday November 9
Partly cloudy with light westerly winds. Eight AM temp −2°, 4 PM −4°. This being the king of England's birthday it has been kept by the steamer as a holiday. I was invited to supper on the steamer, during which the toast was the king's health, and later the president of the United States' health was also given. After supper a dance was in order. It is now midnight and the dance is still going on. Had a pleasant time, the vessel's deck being draped with flags, but did not see the American flag among them. We had our flag flying during the day in honor of England's king.

Thursday November 10
Moderate snow storm from SE. The natives have got their snow and ice houses completed ashore and moved in today. Though the sick woman is unable to be moved [she] is gaining slowly. Two of the native men have gone off deer hunting. Temp 8 AM 6°, 4 PM 26°.

Friday November 11
Moderate storm from NE with snow, increasing at night. Eight AM temp 25°, 4 PM 20°. One rabbit taken today. Mr Ellis in going to his traps found that someone had taken a fox out of one of them, and as one of the steamer's men brought in one last night we think that it was taken from our trap.

Saturday November 12
Partly cloudly but quite pleasant. Wind fresh but becoming light towards

noon. Eight AM temp 2°, 4 PM 2°. Had a dance here this evening. Mr Ellis got a fox in his traps.

Sunday November 13
Partly cloudy. Wind moderate from NW to NE. Temp 8 AM −5°, 4 PM −7°. Captain Bernier read prayers at the grave of the two men who died last April, a number of his crew going and I went with them. One fox taken in trap. Took supper on the steamer.

Monday November 14
Pleasant with light breeze north to NW. Mr Ellis got another fox in his trap. Eight AM temp −18°, 4 PM −16°. Our native who went off a few days ago to hunt deer returned today not having seen any, but brought in six deer which had been taken earlier in the season. The sled sent out by the steamer returned with their deer which they shot.

Tuesday November 15
Pleasant with light northerly winds. Eight AM temp −20°, 4 PM −17°. The sick woman is now able to sit up each day.

Wednesday November 16
Snowing with a strong breeze from SE. Eight AM temp 6°, 4 PM 15°. Major Moodie's son came over and wished me to let him have deer skins enough to have a coat made − he got them. The Major sent over a half-barrel of sugar in exchange for some molasses.

Thursday November 17
Pleasant. Wind light from north to NE, increasing at night. The steamer *Arctic* has rigged up a windmill in order to generate electricity for lighting purposes. This steamer was formerly the German vessel *Gauss*, which was sent on an expedition to the Antarctic a short time ago. Major Moodie and his wife called this forenoon. I was well pleased with her and their call. I sent over two tierces of molasses in exchange for sugar. We have sugar enough for the cabin use but this I wished to get for the men, who could use it in their lime juice. Eight AM temp 22°, 4 PM 10°. Put quite a little snow around the vessel to help keep out the cold.

Friday November 18
Cloudy with fresh breeze from ENE. One rabbit brought in. Temp 8 AM 8°, 4 PM 22°. They are having a dance on board the steamer this evening.

Some of the members of the steamer have begun printing a weekly paper, the first copy coming out this evening. Captain Bernier came over and spent the evening. He is much given to North Pole work and hopes to be able to go on a voyage.

Saturday November 19
Cloudy, easterly weather. Temp 8 AM 18°, 4 PM 19°. We are having very little snow so far this season. We have usually had the vessel well banked in by this time. What snow we have remains soft and is not fit to cut into blocks for building houses. The vessel's leak is now still 300 strokes a day. The sick woman is now able to move around a little. Took a cask out to the reefs and fixed it as a fox trap. Some of the people on board the *Arctic* have printed a small paper this week which it is intended will be published weekly during the winter. A copy of the paper was sent over to me with the request that I would contribute something each week for printing.

Sunday November 20
Cloudy with light breeze from NE and east. Temp 8 AM 13°, 4 PM 11°. Took dinner on board the *Arctic* and this evening we all attended a lecture given by Captain Bernier on different voyages which have been undertaken to the North Pole. I must say that Captain Bernier did extra well. He had many lantern slides in which he illustrated his discourse.

Monday November 21
Partly cloudy, wind light from NE. Temp 8 AM 21°, 4 PM 23°. We have made a new weather vane and put [it] up on our beacon which guides us into the harbor. It is made of tin and represents a whale. Have now commenced to work on ivory and am fixing up a minute glass which was formerly used in recording the speed of a ship. The thickness of the ice today was nineteen inches.

Tuesday November 22
Cloudy weather with light air from NW. Temp 8 AM 21°, 4 PM 20°. Mr Ellis went out to the floe with a native and one man. They got one seal and eleven ducks. Large number of duck remain here wherever they find open water near reefs during the winter.

Wednesday November 23
Cloudy with light wind NW, hauling to NE. Temp 8 AM 12°, 4 PM 18°. The

sick woman is now quite well and has removed to her snow house. Our people who have been off gunning had no success. Our natives which are here are working on their winter clothing and according to their laws cannot hunt seal till finished.

Thursday November 24
Cloudy, wind north light. Temp 8 AM 10°, 4 PM 8°. Captain Bernier and his chief engineer were over to dinner with us as we made a little extra, being Thanksgiving.

Friday November 25
Partly cloudy with wind increasing from north. Temp 8 AM −3°, 4 PM −5°. Opened a cask of bread. We are still unable to get snow to bank the vessel in. A dance is being given on the steamer this evening.

Saturday November 26
Pleasant with moderate breeze from north. Temp 8 AM −8°, 4 PM −10°. Opened a new cask of flour. The bread we opened yesterday is quite buggy. It is bread which was bought from the government after the Spanish war.

Sunday November 27
Pleasant with moderate breeze from NW. Temp 8 AM −21°, 4 PM −19°. Took dinner on the steamer and a Mr Mackean, who is the artist for the expedition, took dinner here. Had the graphophone out this evening.

Monday November 28
Pleasant, wind light from west. Put more of the banking up around the vessel but it is difficult to get snow. Temp 8 AM −21°, 4 PM −10°. Mr Ellis got ten ducks. This evening I took the graphophone down in the forecastle for the benefit of the men. Thickness of the ice $23^1/_2$ inches.

Tuesday November 29
Pleasant though partly cloudy. Temp at 8 AM 2°, 4 PM 0°. Have been at work repairing sleds. Took a walk ashore to the ice pond.

Wednesday November 30
Pleasant with light wind from north. Temp 8 AM 8°, 4 PM 5°. I have been writing a little for the weekly paper. Spent the evening on board the *Arctic*. Had a pleasant time.

Thursday December 1
Partly cloudy with moderate breeze from WNW. Temp 8 AM 5°, 4 PM −2°.
Finished a sled for hauling ice and worked some on ivory. Two of the men
(natives) have gone off after deer skins which were shot last fall.

Friday December 2
Fresh breeze from NW driving the snow. Eight AM temp 6°, 4 PM 6°. The
steamer's people have set many traps for foxes and have taken about
twenty so far.

Saturday December 3
Pleasant with fresh breeze from NW. Temp 8 AM −10°, 4 PM −16°. Had
a dance here this evening. We make a keg of spruce beer each week.

Sunday December 4
Pleasant with a moderate breeze from NW. Temp 8 AM −20°, 4 PM −20°.
Took dinner on the steamer. They are having two meals a day, 9 AM and
4 PM. Major Moodie is now living in his house ashore. They are connected
by telephone. Captain Bernier wh— [whose] mind is on an expedition to
the North Pole and he certainly has a large amount of information on the
subject, and it is very interesting to me in his description of what has
been done and what looks like the best chance of reaching there.

Monday December 5
Pleasant with moderate breeze from NW. Eight AM temp −26°, 4 PM −26°.
Put some more snow up around the vessel but this year we are unable to
get snow enough to bank the vessel and it is therefore much colder in the
vessel than usual. We are not more than half banked in as yet. Thickness
of the ice twenty-three inches.

Tuesday December 6
Pleasant with wind moderate from NNE. Temp 8 AM −23°, 4 PM −21°.
Did some more work on the banking. Mr Ellis got a fox in his traps. Our
two natives returned having gone after the deer skins which were taken
last fall. In order to reach the vessel before dark the skins were left inland
in sight of the vessel. Expect to return for them tomorrow. A sled left the
steamer on a trip to Chesterfield Inlet to carry provision for a sled which
is expected to start tomorrow with Major Moodie's son to go on a trading
trip up Chesterfield Inlet and Baker Lake. This will be likely to take what
trade we might have got away from us. A partridge came close to the
vessel and lit but then continued on south.

Wednesday December 7
Overcast with fresh breeze, increasing north to NNE. Got off two loads of ice. Our two natives returned with the skins they went to get. They brought what they had in skins – thirty-two in all besides three bodies of the deer. They have more deer meat buried, which they may get later in the winter. No deer have been seen during their trip. The wolverines had dug into one of their caches of deer meat and destroyed it. Temp 8 AM $-13°$, 4 PM $-12°$.

Thursday December 8
Pleasant with light air NW to W. Eight AM temp $-11°$, 4 PM $-11°$. Although we have not got banked up as yet with snow we today put the ice up to the windows, thinking it will help keep out the cold. Spent the evening on the *Arctic*. Captain Bernier gave me one of his photographs. A bear was seen out at the floe by Mr Ellis.

Friday December 9
Cloudy with light airs from NE, hauling to south. Temp 8 AM $-5°$, 4 PM $-5°$. A little work is carried on each day. Each day one man has charge of the forecastle and does no outside work but can wash his clothes and care for himself, while another man comes into the house and looks after keeping everything picked up and clean, each man taking his turn. Mr Reynolds has charge of the hole and Brass has charge of keeping the house in order, while the boatsteerers each day keep the water hole cut open so that in case of fire we could get water handy. Then all pieces of tin or glass can be thrown down it and not be liable to cut anyone by stepping on them.

Saturday December 10
Cloudy, partly light airs easterly. Temp 8 AM $-10°$, 4 PM $-12°$. Had a dance this evening. Major Moodie has sent a sled to Baker Lake in charge of his son to trade with the natives. Two sleds have gone, one to return soon but to carry provision for the other. This will take what trade we might have got away from us from that part of the country.

Sunday December 11
Light airs from west and NW. There has been a mist hanging over today. Temp 8 AM $-17°$, 4 PM $-13°$. Mr Ellis got one duck. Captain Bernier and a couple of gentlemen were over here this evening.

Monday December 12
Cloudy with a little snow. Wind NW hauling to north. Temp 8 AM 3°,
5 PM −5°. Major Moodie came over and visited a short while this afternoon and also invited me to call upon him at his house ashore, where he
is now staying. Thickness of ice 23¹/₂ inches.

Tuesday December 13
Pleasant with light winds from NW. Temp 8 AM −23°, 4 PM −23°. Mr
Ellis got one fox.

Wednesday December 14
Very pleasant with light airs from north. Temp 8 AM −25°, 4 PM −22°.
Mr. Ellis got one seal, four ducks, and an owl. Have been over to the
steamer this evening.

Thursday December 15
Very pleasant with light airs from NW. Temp 3 PM −12°. Mr Ellis got ten
ducks and a fox and an owl. We put up a little more snow around the
vessel but there is but little we can get, especially snow we can cut into
blocks. The vessel is much colder than usual because we cannot get snow
to bank with.

Friday December 16
Pleasant with moderate breeze from SE to NE increasing. Temp 8 AM −3°,
4 PM −7°. Having a dance on the steamer. One rabbit taken.

Saturday December 17
Very pleasant with light airs from NW. Temp 8 AM −22°, 4 PM −20°. One
rabbit taken. Captain Bernier came over and spent the evening. Mr Ellis
is preparing to go to Walrus Island tomorrow to camp a few days and try
to get some ducks and seal.

Sunday December 18
Very pleasant with light air from NW. Temp 8 AM −24°. Mr Ellis went
this forenoon to Walrus Island with two natives to remain a few days.
Captain Bernier gave his lecture this evening on the best route to the
North Pole. It was a continuation of the other lecture given a short time
ago. One of Major Moodie's sleds returned this evening which has been
to carry supplies for the sled which was to continue on to Baker Lake to
trade with the natives. They brought back three deer.

Monday December 19
Very pleasant with light airs from north. Temp 8 AM −30°, 4 PM −29°.
The vessel is quite cold in the house, that is to work with comfort. Of
course plenty warm down below. Thickness of the ice thirty-one inches.

Tuesday December 20
Slightly cloudy, light wind north. Got a little more snow around the vessel.
Temp 8 AM −30°, 4 PM −30°. The doctor and the third engineer were
over this afternoon getting items for this week's paper. They are the ones
who typewrite and act as its editors.

Wednesday December 21
Pleasant with moderate breeze from NE. Temp 8 AM −25°, 4 PM——.
Am working on walrus tusks, making a cribbage board. Went over to the
steamer to take what I had written for the weekly paper.

Thursday December 22
Pleasant with light wind north to NW. Temp 8 AM −30°, 4 PM −25°. Got
a little more snow up for banking. Mr Ellis returned from Walrus Island
bringing twenty-six duck. Captain Bernier came over for a visit this evening.

Friday December 23
Partly cloudy with fresh breeze at times from NW. Eight AM temp −14°,
4 PM [−]16°. Got one side of the vessel's banking finished. Today the sun
starts to come north and this we look upon as one of the turning points
of the winter. Two of the people from the steamer were out to the floe
but not returning til 7:30 caused some anxiety, but they came in all right
− Dr Flood and Mr Mackean.

Saturday December 24
Pleasant with moderate breeze from WNW. Temp 8 AM −25°, 4 PM −26°.
Had a dance here this evening.

Sunday December 25
Pleasant with moderate breeze from NW. Temp 8 AM −32°, 4 PM −32°.
I have had presents from some of the steamer's people. The doctor sent
me a case of tobacco in tin boxes called Golden Seal (ten pounds). Mr
Nagle the third engineer sent me four tin boxes of tobacco, golden leaf.
One of the police [sent] a new pipe. Another one [sent] one pound of plug
tobacco. I made each member of our crew a present of tobacco, a towel

and a handkerchief. I called at the major's house but he and his wife were on the steamer so did not see them. Spent the evening on the steamer. Had a very good time but nothing like what we had on the *Neptune* last year. The major was not there.

Monday December 26
Pleasant with moderate breeze from west. Temp 8 AM −32°, 4 PM −35°. Spent the evening on the steamer, they having a dance there. Made a call on Major Moodie and his wife at their house ashore. Thickness of the ice thirty-two inches.

Tuesday December 27
Partly cloudy with moderate breeze from NW, hauling to NE at night. Temp 8 AM −38°, 4 PM −32°. Mr Ellis had intended to go to Walrus Island but it being cold and the prospect of a change in the weather I thought he would do well not to go. A sled from the steamer went – one native and two of the police force. I do not remember of ever having seen the barometer as high as it was this morning – 30.44.

Wednesday December 28
Pleasant with light breeze from NE. Mr Ellis went this morning to Walrus Island to stay a few days hunting. Temp 8 AM −29°, 4 PM −28°. We put more snow up against the vessel. Now we have the sides completed but wish to cover the roof. Major Moodie made a call during the day.

Thursday December 29
Overcast with light winds from ENE. Temp 8 AM −20°, 4 PM −10°.

Friday December 30
Overcast with strong winds from SE to east, driving snow and also snowing. Temp 8 AM −5°, 4 PM −2°. We are eighteen months out today, are all well (the present crew). The vessel's leak is a little less than 300 strokes per day so far this winter. This is our first snow storm this winter.

Saturday December 31
A very stormy day with the wind NE. Temp at 8 AM −5°, 4 PM −9°. Tomorrow starts us on another year and the year in which we go home, and though we have not been successful this year in our voyage I feel a regret in having to let this year pass into history. We have much to be thankful for, myself especially.

6

The Second Winter
(1 January – 9 May 1905)

Sunday January 1
A stormy day with driving snow from NE. Temp 8 AM −23°, 4 PM——.
Took dinner on the steamer with Captain Bernier. Now we start upon a
new year and the one in which we expect to return home.

Monday January 2
Weather improving. Wind north moderate. Temp 8 AM [−]19° 4 PM——.
Mr Ellis returned, has taken four foxes. A dance is going on on board the
steamer. Thickness of ice 33¹/₂ inches.

Tuesday January 3
Very pleasant with light airs from NW. Temp 8 AM −22°, 4 PM −20°. We
have today finished banking the vessel in. This work we have usually had
done by the twentieth of November but this winter we have not had the
snow to do with. Major Moodie was over awhile this afternoon.

Wednesday January 4
A very pleasant day. Light air north to NW. Temp 8 AM −24°, 4 PM −21°.
Went out to the floe edge with Mr Ellis. We got eight ducks, the distance
out being about six miles. There were many ducks but we could not get
near them on account of so much young ice. No seal to be seen. Captain
Bernier and the artist were over this evening. One of our boatsteerers,
Manuel, is now sick and has the doctor's attention from the steamer – a
cold on his lungs.

Thursday January 5
Cloudy but mild weather, light air north. It seems warm to us. Temp

8 AM −12°, 4 PM −5°. Our sled has gone to get deer meat and will stop at Whale Point and leave a barrel of bread so that any one of our natives coming from the north may find it in case of need. We weighed a white fox which tipped the scales at six pounds four ounces. A blue fox weighed four pounds nine ounces.

Friday January 6
Partly cloudy with light air from north. Temp 8 AM −14°, 4 PM −6°. Major Moodie's son, who has been gone on a trading trip to Chesterfield Inlet and Baker Lake, returned last evening. Two native (men) returned with his party. What success he has had is not made known outside. Mr Ellis and I spent the evening on the steamer, it being known as Little Christmas.

Saturday January 7
Partly cloudy. Wind north light. Temp 8 AM −20°, 4 PM [−]21°. Our two men who are sick are on the gain. Dr Flood of the steamer *Arctic* attends them. In one of the native families here there was some medicine which was poison if taken internally. Two of the children got hold of it last evening and drank some of it after putting it in water. It made them sick and caused them to vomit but it nearly cost one of them his life. Had a dance here this evening.

Sunday January 8
Very pleasant with light winds from north. Temp 8 AM −23°, 4 PM −18°. Our sick men are gaining. I took supper on the steamer while the doctor and Mr Mackean took supper with the schooner. One rabbit shot.

Monday January 9
Pleasant with light westerly winds. The steamer people (police, four) started off on a salmon trip for a few days. Temp 8 AM −21°, 4 PM −19°. Spent the evening at Major Moodie's house, Mrs Moodie being very hospitable. Our sick men were up around today. Thickness of the ice 35¹/₂ inches.

Tuesday January 10
Very pleasant with light westerly winds. Temp 8 AM −17°, 4 PM −13°. I have been working carving out some designs on walrus tusks and took them over to show Mrs Moodie, who is interested in such work as she is at work carving woodwork such as trays. I gave her a walrus tusk, which pleased her very much. She is going to try her hand at carving some

designs on it. Had a pleasant evening, they doing all they could to make
it pleasant.

Wednesday January 11
Very pleasant with light airs from NW. Temp 8 AM $-30°$, 4 PM $-24°$.
Today one of the boatsteerers refused to do some work required of him
and I had him remove to the forecastle. His name is Charles Tucker. I
am also finding that my steward is taking [some] of the small stores for
his own use as trade to others.

This is my wife's birthday and the anniversary of our marriage, and
last night in speaking of it to Mrs Moodie she told others of it and Major
Moodie sent a note with a box of candy, while this evening Captain Bernier
with Mr Vanasse came over and invited us over to lunch, where wine
was passed, and in the mixup it became known as my birthday. Mrs
Comer's health was drunk by all.

Thursday January 12
Very pleasant with light winds from NW. Temp 8 AM $-32°$, 4 PM $-29°$.
Sometime ago my chair (office chair) broke down and I took it to the
steamer thinking that they might make a better job in repairing than I
should. It was finished today. I was quite disappoined in the work.

Friday January 13
Pleasant with light winds NW. Temp 8 AM $-24°$, 4 PM $-17°$. Our natives
(two) who went off to get some deer meat and leave a barrel of bread at
Whale Point returned at 2 AM bringing two deer, one of which they shot.
They saw quite a number but being light airs and calms could not approach
them. Mr Ellis and the artist of the steamer are preparing to start off
tomorrow for Walrus Island to camp a few days hunting seal and ducks.
They will take two young natives with them.

Saturday January 14
Partly cloudy. Wind increasing and driving the snow, which fell in considerable quantity last night. Mr Ellis and the artist Mr Mackean with
two native boys started for Walrus Island this morning to be gone several
days. Temp 8 AM $-1°$, 4 PM $-4°$. Had a dance this evening. Some of the
natives of the steamer came up from a hunting trip from Walrus Island,
having taken ten seal. They had a party out fishing for four days. They
got six salmon and several bad colds.

Sunday January 15
Partly cloudy with moderate breeze from NW. Temp 8 AM −21°, 4 PM −25°. Had Mr Moodie (the Major's son) and the third engineer come and take dinner with us. I went over and spent the evening.

Monday January 16
Partly cloudy, wind NW. Temp 8 AM −31°, 4 PM [−] 27°. I am still working upon walrus ivory, making cribbage boards. The vessel is now leaking a little more, about 500 strokes a day. Thickness of the ice thirty-nine inches.

Tuesday January 17
Pleasant with light wind west, and at night hauling to NE, becoming overcast. Temp 8 AM −31°, 4 PM −27°. Major Moodie was over a short while this afternoon. He spoke as though he would like to purchase our house lumber in the spring when we were ready to take it down.

Wednesday January 18
Moderate storm from SE, then backing to north. Temp 8 AM −10°, 4 PM −9°. Such weather seems warm to us. Our sled came back from Walrus Island with one of the natives and Mr Mackean, who seems to have had enough of camping out, Mr Ellis and the native boy (Mike) remaining at the island. They have taken four ducks.

Thursday January 19
Strong winds from north driving the snow. Temp 8 AM −28°, 4 PM −25°. One of the men being sick (indigestion) I sent for the doctor who is doing what is necessary (Charles Tucker). Made a call at the steamer this evening.

Friday January 20
Blowing strong this morning from north but becoming better weather later. Temp 8 AM −25°, 4 PM −23°. The vessel is now leaking 1,200 strokes per day. The sick man (Tucker) is now quite well. Captain Bernier and a party returned from a four day fishing trip – nothing taken. A dance on the steamer this evening. I spent part of the evening there.

Saturday January 21
Very pleasant with light winds NW. Temp 8 AM −27°, 4 PM −25°. Spent the evening on the steamer, it being the birthday of two of the saloon party, which had to be celebrated with a glass of wine each. Opened a

barrel of pork. Our sled went this morning to get some deer meat near Whale Point.

Sunday January 22
Pleasant with fresh breeze from NW. Temp 8 AM − 20°, 4 PM [−]24°. Took dinner on the steamer while Mr Pelletier (Inspector of Police) with Mr Mackean took dinner here with Mr Reynolds. The steamer people seem to enjoy coming here to dinner.

Monday January 23
Very pleasant with light wind north. Temp 8 AM − 28°, 4 PM − 24°. Mr Berg of the steamer (mate) took dinner with us. The days are getting much longer. By this I mean that it is quite light at 4 PM so that we can see in the house. Our sled returned with about 300 pounds of deer meat which it had gone to get. No deer seen but a number of tracks were seen. Thickness of the ice today was forty-three inches.

Tuesday January 24
Partly cloudy with moderate breeze from west. Temp 8 AM − 23°, 4 PM − 16°. Our sled started again for Walrus Island taking some provision for Mr Ellis, who has been down there for a few days, and bring him back if he wishes to come.

Wednesday January 25
Partly cloudy with the wind moderate from NNE. Temp 8 AM − 14°, 4 PM − 14°. Captain Bernier was over and spent the evening. He made me a present of a magnifying glass, a good large one.

Thursday January 26
Very pleasant. A series of games were held today on the ice by both vessels' crews and the police force. One in which I took part was a 100-yard dash running with Captain Bernier, in which I came in ahead but as he is some six years older and quite fleshy is not to be wondered at. The day passed very pleasantly and all enjoyed themselves. Light wind from NW. Temp 8 AM − 29°, 4 PM − 26°. Two sleds came this afternoon to the steamer with trade. There were six men in the party.

Friday January 27
Pleasant with light NW wind. Temp 8 AM − 21°, 4 PM − 25°. I was presented with a very pretty metal cup as a prize for winning the foot race yesterday.

Saturday January 28
Pleasant with light to strong breeze NW. Temp 8 AM −27°, 4 PM −27°.
Have taken plaster casts of the natives' faces who came up and are
stopping at the steamer. They are from the SE of Chesterfield Inlet and
their tribe is known as the As-shock-miut or Show-vock-tow-miut. There
are seven men and two women in the party and have been six weeks
coming.

Sunday January 29
Very pleasant with moderate NE winds. Temp 8 AM −33°, 4 PM −30°.
Took dinner on the steamer. Mr Ellis returned from Walrus Island – five
foxes and ten ducks, one seal skin. Had the graphophone out this evening
for the natives and got a couple of records of their songs. Took the census
of the Show-vock-tow tribe which live near Baker Lake to the SW. They
numbered 178 all told. Also took a photograph of three of the men.

Monday January 30
Pleasant with moderate breeze from NNE. Temp 8 AM −20°, 4 PM −17°.
Have been taking plaster casts of some of the Show-vock-tow-miut. Took
casts of their faces and some of their hands, three men and two women.
Some of the steamer's people came over to see how it was done. Thickness
of the ice forty-four inches. Gave the men (our crew) their monthly supply
of tobacco and things that they need. The barometer is unusually high,
it being 30.74 this 9 PM.

Tuesday January 31
Overcast with light winds NNE. Temp 8 AM −17°, 4 PM −12°. A seal was
taken by one of our young natives out at the floe and brought to the vessel,
then the young man John L., according to custom, put a little fresh water
on the outside of the window for the seal's spirit to drink. This is a custom
among them. Filled up the plaster casts.

Wednesday February 1
Partly cloudy with light winds from NW and west. Temp 8 AM −20°, 4 PM
−22°. Took some letters over to the steamer which the men have written
home, as Major Moodie expects to have a sled start for Fort Churchill
on Saturday, February 4. Two white men and two natives are to go while
another sled will go part way to carry provision for them. One of our
natives has begun building a snow house near the vessel. Having finished
working on deer skins they can now live on the ice. I agreed with Major

Moodie to let him have our lumber in the spring for $25.00 per thousand feet. They are having a dance on the steamer. Took a picture of a group of the southern visitors. Barometer 30.90 – this is higher than I have ever known it to be.

Thursday February 2
Partly cloudly with moderate breeze from west. Eight AM temp −23°, 4 PM −20°. Had a singing concert this evening on the steamer. Have been developing pictures and it is now 2 AM.

Friday February 3
Very pleasant with light winds from west. Temp 8 AM −21°, 4 PM −21°. Mr Ellis returned to Walrus Island taking with him the two native boys Mike and John L., the carpenter also going for the outing. Some more natives came to the steamer from the southward to trade. Took some pictures of Mr Moodie and the corporal McArthur, who are to start tomorrow for Fort Churchill to carry the mail. Spent the evening at the major's house ashore.

Saturday February 4
Very pleasant with light winds from WNW. Temp 8 AM −27°, 4 PM −23°. The party carrying the mail to Fort Churchill left at 11 AM, consisting of three sleds, six men and thirty-two dogs, one sled to return and two men, after going part way.[1] Got off a cask of bread and opened. Some more natives came to the steamer this evening.

Sunday February 5
Partly cloudy. Wind light from NW. Temp 8 AM −21°, 4 PM————. Took dinner on the steamer.

1 About 200 personal letters and some official reports composed the mail, which was carried to Churchill by a party consisting of Corporal McArthur, A.D. Moodie, Harry Ford the interpreter, and the Eskimo Tupearlock, or 'Tupik.' They took two sleds and twenty-four dogs and were accompanied for a distance by a third sled to share the load. There were difficulties on the way south and several dogs were lost, but the party reached Churchill, handed over the mail to the Hudson's Bay Company, and returned to Fullerton Harbour by 10 April, having completed a journey of about 1,100 miles in two months. Employees of the Hudson's Bay Company carried the mail on to Winnipeg, and it eventually reached Ottawa 'in a bad condition from having been under water while crossing Hudson Bay.' The letters were dried out and forwarded to their various destinations ('Mail to & from Hudson Bay, via Fort Churchill 1904–5,' RCMP Records, vol 314, file 188).

Monday February 6
Partly cloudy with light NW winds. Temp 8 AM −28°, 4 PM −23°. The thickness of the ice was fifty-one inches. Major Moodie gave an exhibition of lantern slides this evening. It had the appearance to me as though he wished to impress the natives on the power and greatness of Canada.

Tuesday February 7
Overcast with wind SE, backing to east and increasing with snow falling. Temp 8 AM −13°, 4 PM −6°. The vessel is leaking still at the rate of 1,500 strokes a day. The other family of natives have moved off near the schooner (Jimmy). I am still at work making cribbage boards out of walrus tusks – have finished four and expect to make two more.

Wednesday February 8
A stormy day, wind NE driving the snow. The doctor, Mr Nagle and Mr Mackean were over today for a call. This evening a dance on the steamer. I have been writing a piece for our paper, which is to appear once in two weeks hereafter.

Thursday February 9
Stormy day. Wind heavy from NNE driving the snow. Temp 8 AM −22°, 4 PM −21°. Two natives who had gone as extra men with the mail sleds returned today. One of them being sick was the direct cause of their returning. He left the party at Chesterfield Inlet.

Friday February 10
Stormy with wind NNE driving the snow, not snowing but driving the surface snow. The sled came up from Walrus Island – one native boy John L. and the carpenter. Brought one wolverine and one fox. I have five cribbage boards done. The vessel's leak is still 1,200 strokes a day. Temp 8 AM −22°, 4 PM −20°.

Saturday February 11
Blowing strong from NNE, quite clear overhead. Temp 8 AM −16°, 4 PM −11°. Opened a new cask of flour. Captain Bernier was over and spent the evening. Got out sugar.

Sunday February 12
Partly cloudy. Wind hauling from NNE to SE, and becoming overcast. Temp 8 AM −11°, 4 PM −4°. Mr Mackean took supper with us. This evening I

spent part of on the steamer. One of our men becoming sick, I was called back.

Monday February 13
Very pleasant, light NW air. Temp 8 AM −21°, 4 PM −11°. The sled was started back to Mr Ellis this morning. A number of natives who came to trade with the steamer a short time ago left here today to return. Our sick man is much better today. The doctor calls it muscular rheumatism. He came over and left some medicine for him. Major Moodie was over for a call. The ice measured 48¹/₂ inches.

Tuesday February 14
Very pleasant with light airs from NW. Temp 8 AM −15°, 4 PM −10°. One of the native women went off fishing. She brought back two salmon trout. One of the men got a fox in his trap.

Wednesday February 15
Partly overcast, wind moderate from west. Temp 8 AM −17°, 4 PM −5°. A little snow falling towards night. A dance on the steamer this evening. Dr Flood was over awhile this afternoon.

Thursday February 16
Partly cloudy, wind light from NW. Temp 8 AM −17°, 4 PM −20°. This day was given up to games, one in which I ran a race with Captain Bernier, he doing a little better than last time but still second. This evening a concert on the steamer, where the Major presided. Mrs Moodie was also present.

Friday February 17
Very pleasant with moderate breeze NW. Temp 8 AM −33°, 4 PM −29°. Today we filled the steerage run with coal from the forward run in the cabin and took an inventory of canned goods. Captain Bernier was over for awhile this evening and at 9:30 three sleds with our natives arrived from Repulse Bay – Harry, Sam and Melichi, Billy, Tom Nolyer, and all their families. They left Repluse Bay about six weeks ago and have had good success in hunting.

Saturday February 18
Very pleasant with moderate NW winds. Temp 8 AM −37°, 4 PM −20°. The boatsteerer who some time ago refused to do what I wished and who

had been placed in the forecastle having apologized, I allowed him to return to the steerage. I did what I thought was right to encourage him to do so, as he had always been a good man and I did not wish to see him put down. A dance here this evening. The natives brought in $3^1/_2$ saddles of deer meat.

Sunday February 19
Very pleasant with moderate wind from NW. Temp 8 AM $-28°$, 4 PM $-28°$. Mr Ellis returned from Walrus Island, had taken two seal and one fox. I took supper on the steamer this evening. While I was in one of our native igloos one of the natives went through one of their performances. He being first made apparently securely fast with a line, he set himself free, and in doing this it was claimed that his spirit left the body and went to the spirit world and played football with others, the performance lasting about a half an hour.

Monday February 20
Pleasant, wind NW fresh breeze. Temp 8 AM $-25°$, 4 PM $-18°$. Some of the natives started off to get some part of their load which they had left near Whale Point. Today we put in a part of a plank in the cabin from one of them having become rotten. Our carpenter was quite sick this evening but after having the doctor come over and treating him he is now better – something he could not digest. Thickness of the ice fifty-two inches.

Tuesday February 21
Partly cloudy with light NW wind. Temp 8 AM $-32°$, 4 PM $-23°$. The sleds returned from having gone after the extra goods which the natives had left near Whale Point. Our sick man is quite well today. Three and a half saddles of deer meat brought in.

Wednesday February 22
Cloudy with wind east light. Temp 8 AM $-17°$, 4 PM $-16°$. Our natives went off walrus hunting. They got one and returned by 4 PM. A dance on the steamer. I am quite sick with a cold and have to remain in bed.

Thursday February 23
Pleasant with light NE wind. Temp 8 AM $-18°$, 4 PM $-17°$. Can sit up this afternoon. The doctor and Captain Bernier have both been over, Captain Bernier sending over a bottle of wine.

Friday February 24
A stormy day, wind strong from east. Temp 8 AM −7°, 4 PM −6°. My cold may be a little better but still I cannot speak aloud. The vessel's leak is now reduced to about 400 strokes a day.

Saturday February 25
Weather improving. Wind moderate from north NE. Temp 8 AM −9°, 4 PM −8°. A dance this evening on the schooner. Major Moodie sent off a native requesting the use of five or six dogs for a trip to the south of Chesterfield Inlet. I sent back a note telling him the conditions he could have them on.

Sunday February 26
Very pleasant with light wind from NNE. Temp 8 AM −17°, 4 PM −9°. In reply to my note to Major Moodie he returned a very sarcastic answer that he would get along with what dogs he had. Our natives got a seal at the floe today. Mr Ellis and the doctor of the steamer are going with a party of our natives to Walrus Island to remain a few days. Some more natives came to trade with the steamer, from Baker Lake.

Monday February 27
Very pleasant with winds light from NW to NNW. Temp 8 AM −21°, 4 PM [−]15°. Mr Ellis and the doctor went off this morning with a party of natives to Walrus Island to stay a few days. The rest of the natives went out [to] the floe but were unsuccessful. Thickness of the ice fifty-seven inches.

Tuesday February 28
A little cloudy, wind variable. Temp 8 AM −21°, 4 PM −5°. My cold is a little better but still I cannot speak aloud yet and cough a great deal, and can get very little sleep. Mr Mackean and Mr Nagle were over this afternoon and I invited them to come to dinner tomorrow and try some walrus meat cooked up.

Wednesday March 1
It has become a very stormy day with NE gale, snow. Temp 8 AM 1°, 4 PM 0°. Mr Mackean and Mr Nagle were over to dinner with us 4 PM as we only have two meals a day and like it much better than having three. My cold is improving.

Thursday March 2
Very pleasant with light NW winds. Temp 8 AM −22°, 4 PM −20°. The storm was so great last night that while our carpenter was attempting to come from the steamer after the dance was over he lost his way and brought up on the island astern of us.[2] When he found he was lost he made a hole in the shore ice and stayed till this morning, when he came to the vessel. Another man after finding the vessel could not seem to find the gangway but made an attempt to get on board by the way of one of the openings in the banking for the window, but finally got in all right. My cold seems a little better but still not able to speak very loud.

Friday March 3
Pleasant with moderate breeze from NW. Temp 8 AM −24°, 4 PM −17°. I am feeling quite better as my cough is more loose. Took some pictures of our natives. The light in the house seems to be extra good for such work.

Saturday March 4
Very pleasant with light winds from NW. Temp 8 AM −26°, 4 PM [−]15°. My cold is on the improve so that I went out to the igloos for a few minutes. There was a dance this evening here. Captain Bernier was over and spent the evening and later sent over a bottle of wine. Mr Mackean was over this afternoon. Opened a barrel of pork. A sled started today from the steamer to place provision for the return party from Fort Churchill.

Sunday March 5
Very pleasant with light NW breeze. Temp 8 AM −28°, 4 PM −18°. Our natives got two seal at the floe. Mr Ellis sent up for more provision. They have taken one wolverine and one fox. The sled will return tomorrow. Dr Flood of the steamer wishes to stay four days longer.

Monday March 6
Very pleasant with light NW winds. Temp 8 AM −27°, 4 PM [−]16°. Thickness of the ice fifty-eight inches. The sled returned to Walrus Island taking

2 Blowing snow constitutes a severe hazard in arctic regions. In the autumn of 1904 a
  trail marked with tall sticks had been laid out over the ice between the steamer and
  the police post. Vanasse called it l'Avenue des Dames, for reasons which we can
  only imagine (Vanasse, 'Relation sommaire du voyage de l'Arctic, 1904–5,' 35). It
  appears that the traffic between the Arctic and the Era was not heavy enough to
  warrant a marked trail and its nature was not such to deserve a similar appellation.

more provision. Two duck brought in today. Have had the carpenter start in making a new boat mast for my boat.

Tuesday March 7
Very pleasant with light NW winds. Temp 8 AM −29° 4 PM [−]24°. We cut the rudder clear of ice so that in case the vessel should come up, which she will do later, she will not damage the rudder. One of the native women at the steamer gave birth to a child, a girl. One family of the natives who went to Walrus Island with Mr Ellis and the doctor returned today, one of their children being sick. This evening the natives are having an anti-cooting performance to drive away the evil spirit which is supposed to be the cause of the child's sickness. The Eskimos have no medicines of any kind. Some natives who came to the steamer a few days ago tell of one of their number who was dying, who before he stopped breathing rolled him [self] up in deer skins for burial. He was a person I knew well, known as Seco (which means ice).

Wednesday March 8
Very pleasant with light north wind. Temp 8 AM −29°, 4 PM −24°. Have been taking pictures, using Captain Bernier's camera and plates, he to have half. The sled returned to Walrus Island taking another family of natives. My cold is much better.

Thursday March 9
Overcast with moderate breeze from west. Temp 8 AM −30°, 4 PM −10°. Let Captain Bernier have three dozen lamp chimneys and a few burners. Got out another boat mast for Mr Reynolds. Mr Nagle was over, third engineer, and stayed awhile this afternoon.

Friday March 10
Pleasant with wind moderate from NE. Temp 8 AM −21°, 4 PM −12°. Took a couple of pictures of the natives in their snow houses. Some of the native men were in the cabin this evening and I entertained them by telling of my trips after seal and sea elephant and a description of those countries. All natives have the freedom of the cabin to come and go when they like.

Saturday March 11
Pleasant with light NE winds. Temp 8 AM −8°, 4 PM −3°. A dance this evening here. Captain Bernier was over and spent the evening. The natives

go out to the edge of the floe each day but have no success in sealing. The track of bears are occasionally seen.

Sunday March 12

Pleasant at first but becoming overcast at night. Wind west. Temp 8 AM −14°, 4 PM −7°. Took a number of pictures of the natives with some of them especially painted to show the tattooing of the different tribes, and while I was developing them word came that our carpenter, while playing football, had got his leg broken just above the ankle, both bones. I then went over to the steamer and had him brought over, Sergeant Hayne of the police force attending to the fracture and doing the work with much skill. We had the carpenter placed in the steerage and he is now as comfortable as can be under the conditions. One rabbit taken.

Monday March 13

Very pleasant with light NW winds. Temp 8 AM −15°, 4 PM −10°. The sled came up from Walrus Island, the doctor returning to the steamer. He came over this afternoon with Sergeant Hayne to see the carpenter's leg. He is doing as well as can be expected. Two rabbits were taken today. Mr Ellis sent up one fox. They have taken two seal in all since they have been down there this time, something over two weeks. Thickness of ice fifty-six inches.

Tuesday March 14

Very pleasant with light NW wind. Temp at 8 AM −21°, 4 PM −12°. Though the doctor is at the steamer Sergeant Hayne continues to look out for the carpenter's broken leg, coming over morning and evening. Captain Bernier was over and spent the evening.

Wednesday March 15

Very pleasant with light winds NW to SW at night. Temp 8 AM −25°, 4 PM −15°. Sergeant Hayne comes over to look after the carpenter's leg, which is doing as well as can be expected. The doctor does not come near.

Thursday March 16

Very pleasant with light NW winds. Temp 8 AM −21°, 4 PM −18°. Mr Ellis returned from Walrus Island having lost his boat, the ice breaking off and taking it with it. They have shot several seal but they would sink and could not be got. One rabbit shot by one of the native boys.

Friday March 17
Very pleasant with light NW winds. Temp 8 AM −29°, 4 PM −24°. This morning I got up at five o'clock, took a couple of pictures – one of the barracks and one of the schooner – thinking to get the best contrast at sunrise, the temperature at the time being −35°. A dance on the steamer this evening.

Saturday March 18
Cloudy with a light fall of snow. Temp 8 AM −24°, 4 PM −10°. Some more of our natives arrived today from Repulse Bay, Beach Point – Gilbert and Albert with their wives and two children. This evening a sled arrived from Chesterfield Inlet to trade, two natives stopping at the steamer and two coming here to trade. Those that went to the steamer brought a letter from a ship which is making the Northwest Passage. *Gjoa* is the way her name is spelt, Captain R. Amundsen.[3] This letter, it is reported, leaves them all well, a crew of seven men, and they were in lat 68° 38′ north, long 96° west, frozen in. The letter left them October 15 1904. The letter has been brought by a native known as Blockhead.[4] They have accom-

3 The first great achievement of the Norwegian explorer Roald Amundsen (the man who was to beat Scott to the South Pole in 1911, attempt the northeast passage in 1918, and fly over the North Pole in a dirigible in 1926), was the completion of the northwest passage. He began the journey in the small sloop *Gjoa* in 1903, reached King William Island and wintered there for two successive years, continued on in 1905 to the mouth of the Mackenzie River, wintered at King Point, and reached San Francisco in 1906. This was the first time anyone had travelled by sea between the Atlantic and Pacific oceans north of the North American mainland.

4 Blockhead, or Artungelar (variously spelled) had left the *Gjoa* at King William Island on 28 November 1904 (not 15 October, as Comer states), bearing letters for the ships at Fullerton Harbour. He was an Eskimo who had guided white explorers (including David Hanbury) and who had seen something of the world beyond the tundra. Once he had visited Winnipeg and 'become acquainted with all the latest discoveries such as the telephone, railways [komatiks on wheels of iron pulled by iron dogs, he told Vanasse], electric light, and – whisky.' Amundsen, on their first meeting, 'tried to explain that the teetotal movement was the latest advance in the region, but he would not listen to it. At last he asked straight out for some spirits ...' (Roald Amundsen, *The North West Passage* [London: Archibald Constable 1908], 264–7). Artungelar's journey to Hudson Bay appears to have been roundabout, for it took almost four months. It was also difficult. He started with one companion and only four dogs, three of which died en route. For shelter he had only a miserable piece of cloth. The accidental discharge of a gun severely wounded his hand (Vanasse, 'Relation sommaire,' 77). However, after a week or so at Fullerton Harbour he was ready to head back towards King William Island with mail from Comer, Bernier, and Moodie and ten extra sled dogs which he delivered to Amundsen on 20 May.

plished their work and expect next summer to push on towards Bering Strait. The man is to return with an answer. The letter was to the commander of ships at Cape Fullerton. Sergeant Hayne was unable to attend to the doctoring of the carpenter so Doctor Flood came over this evening. St Patrick's Day was celebrated in true Irish style on the steamer from the major down.[5]

Sunday March 19
Very pleasant with light NW winds. Temp 8 AM −20°, 4 PM −10°. We now are feeding forty-four people besides our crew of seventeen and we are not getting any fresh meat from the natives, though we did get three saddles from the sled from Chesterfield Inlet. Had the graphophone out this evening and got two more records of native songs. From the Kenepetu sled we got twenty-two fox skins, two wolf and three bear skins, also two wolverine. I got a knife of native copper.

Monday March 20
Very pleasant with wind light from NW. Temp 8 AM −12°, 4 PM −4°. This evening I received a note from Major Moodie regarding the Norwegian

5 Moodie, whose name bespeaks an Irish ancestry, was enthusiastic about the celebration of Saint Patrick's Day. In 1904, 17 March had been declared a holiday for the *Neptune* crew, a football match was played in the afternoon and a dance took place in the evening (Moodie, 'Copy of Daily Diary,' 56). Moodie declared a holiday in 1905 as well; there was a special issue of rum to the men, and the officers drank toasts in champagne. Bernier, although French Canadian, joined in happily, hinting cheerfully of Irish blood somewhere in the family background (Vanasse, 'Relation sommaire,' 75). Three months later Bernier, on behalf of the French Canadians among the crew, proposed to Moodie that they organize a celebration for Saint Jean Baptiste Day (24 June), the 'grande fête dans la province de Québec' – 'leur fête national.' Because it would fall on a Saturday, normally a holiday anyway, no inconvenience would be caused. But Moodie would not allow any celebration: 'There is only one holiday in this country and that is the King's birthday' (ibid. 122). When 24 June arrived, Vanasse noted disconsolately, 'Notre fête nationale passera inaperçu. Les Canadiens français de Fullerton sont moins chanceux que leurs amis Irlandais, dont la fête a été célébrée avec le plus d'éclat possible à l'Arctic' (ibid. 123). It was later alleged by a crew member of the *Arctic* that Moodie had discriminated against French Canadians by punishing them more severely than others and by refusing to allow fish on Fridays. Moodie assured the comptroller of the Royal North-West Mounted Police that he was not antagonistic towards French Canadians but on the contrary friendly (Moodie to White, 30 June 1906, RCMP Records, vol 320, file 543). Bernier later confided to a reporter, 'I don't like to speak disparagingly of anyone, but I must say that Major Moody [*sic*] was an impossible man to work with' (*Montreal Star*, 10 October 1904, RCMP Records, vol 307, file 74).

ship from which he had a letter saying that her captain would like very much to have eight dogs sent to him, and the major not being able to send them requested to know if I would do so, he meaning to buy them. I wrote back saying I would send the dogs but there should be no pay for them.

Tuesday March 21
Very pleasant with light winds from NW. Temp 8 AM −19°, 4 PM −12°. I took plaster casts of the faces and hands of three of the natives who came last [Saturday]. One of them had his hand badly shattered by a bullet. His hand I did not take. The accident happened about three days before he arrived here. His name is Blockhead and his native name is Artung-e-lar. He is the one who has brought the letter from the Norwegian explorer and is to return. I received a note from (I suppose) Major Moodie, though it was not signed, declining my offer to take the dogs as a gift to the Norwegian captain. I shall write to the captain telling him what I should be pleased to do but that he will have to take the will for the deed. It looks as though Major Moodie wants the honor or glory of furnishing the dogs but I do not intend to have it said that I had to be paid in order to help a person in need − that is not American. I think it would choke the major to say the word American as he always speaks of the United States people and directs letters to me as captain of the U.S. schooner *Era*.

Wednesday March 22
Overcast with wind hauling from NE to south. Temp 8 AM −19°, 4 PM +12°. Our natives went walrus hunting and had the good luck to get three and one seal. They got back at 9 PM. A little snow falling. Took a couple of pictures of the natives, one of a group of women (arctic belles) and one of Artung-e-lar.

Thursday March 23
Cloudy with wind light from NW to NE, a little snow falling. Temp 8 AM +16°, 4 PM +12°. I have written a letter to the captain (Amundsen) of the exploring vessel *Gjoa* which the native Artung-e-lar will take with him when he returns.

Friday March 24
Overcast with light NNE [wind] hauling to ENE with a little snow falling. Temp 8 AM −7°, 4 PM +8°. Have made arrangements with Major Moodie to the effect that he can buy five dogs of my natives and send to the

Norwegian captain and that I would also send five, so in this way we may both share in the pleasure of helping him.

Saturday March 25
Overcast with wind light from NE to SE and south, a little snow falling. Temp 8 AM +18°, 4 PM +23°. Our natives went out walrus hunting and returned at 10 PM with a large one, or part of it. Will have to go tomorrow for the rest of it, having left it on the shore ice. The weather being thick I went out to meet the sleds, taking a lantern with me thinking to help them find the vessel. Have written a letter and gave it to Captain Bernier to send with his mail to the Norwegian vessel. Filled up the plaster cast which I had taken.

Sunday March 26
Very pleasant with wind light from NW but at night working to NE. Temp at 8 AM 7°, 4 PM 14°. A sled went out and got the remainder of the walrus meat. Artung-e-lar the native and his party got away to go to Baker Lake and from there to the Norwegian vessel. The major sent over here for the ten dogs, five of which he will pay my natives for and the other five which I have sent as a gift to the captain of the expedition for his vessel's use. The man expects to reach there the first of June.

Monday March 27
Overcast with light NE to east winds. Temp 8 AM 3°, 4 PM 7°. Thickness of the ice seventy-two inches. Our natives who let Major Moodie have the dogs (five) that he might have the pleasure of sending half of the team went to the barrack and got what he called pay for them. To one (Harry) he gave a shirt, a cheap pair of pants, and a handkerchief. To Melichi the same. To Gilbert one pair cheap pants, a jack knife, and about six yards calico. To Ben two handkerchiefs, one box of primers, one cap, and a cheap watch chain. To Sam a two-quart tin pail, a box of primers and a pair of overalls. The whole is not the value of a single dog, especially where a government is buying them for another.[6] Natives with three sleds

---

6 Moodie was critical of the trade values offered by whalemen (see Introduction n 46) but here, judging from Comer's remarks, he himself was offering less than a whaling captain would have given. In general his attitude towards Eskimo trade does not appear to have been inspired by imagination or lofty ideals; he advised the government to send north coloured beads, 'cheap jewelry,' and 'remnants of cheap gaudy prints' for dresses, adding that 'the natives are asking for towels, soap, small mirrors &c' (Moodie to White, 9 December 1903, p 8, RCMP Records, vol 281, file 716).

have gone off deer hunting, one towards Depot Island, the other towards Whale Point and Yellow Bluff.

Tuesday March 28
Overcast with light wind from east. Temp 8 AM 16°, 4 PM 25°. Each day I work but at night can't see as I have accomplished but little. All the natives seem to have an itch. Our carpenter is doing well. The vessel leaks about 500 strokes a day. Mr Mackean and the purser Mr Weeks were over to supper, walrus meat being the principal thing.

Wednesday March 29
Overcast with light east winds, with a little snow falling. Temp 8 AM 23°, 4 PM 23°. A dance on the steamer this evening. I went over and called on Captain Bernier. The first snow bird seen today.

Thursday March 30
The weather clearing this afternoon and wind backing from east to north. Temp 8 AM 13°, 4 PM 7°. Mr Mackean, the artist of the steamer, painted a water color sketch of the harbor and schooner, finished it this evening. It looks very nice. Some of my natives have been making ivory carving for me lately.

Friday March 31
Very pleasant with light NW winds. Temp 8 AM 3°, 4 PM 7°. This weather seems warm to us, so much so that we do not have a fire in the cabin.

Saturday April 1
Weather becoming overcast with wind backing from NW to SE. Temp 8 AM −2°, 4 PM 13°. Mr Ellis and some of the native boys went off after salmon but did not get any.

Sunday April 2
Overcast with light fall of snow, wind light from ESE. Captain Bernier and the chief engineer were over to dinner to eat walrus meat. Temp 8 AM 13°, 4 PM 18°. One of the three parties which went away a few days ago returned this evening with seven deer and two young bear cubs. They shot the old bear but lost it in the water and the tide carried it away.

Monday April 3
Quite pleasant with light airs, variable. Temp 8 AM 18°, 4 PM 27°. Took

away the banking and cut the stem clear of the ice. Did not take the banking away down to the ice.

Tuesday April 4
Partly cloudy, wind light from NW. Temp 8 AM 19°, 4 PM 27°. Got the boats off, five of them, the sixth one being up towards Whale Point. Took a picture of the carvings which the natives have made out of walrus tusks. Captain Bernier was over awhile this evening.

Wednesday April 5
Very pleasant with a moderate breeze from N to NE. Temp 8 AM 6°, 4 PM 11°. We are working on our boats. Another one of our sleds returned from deer hunting bringing seven deer (Jimmy). The vessel is coming up slowly by the stern but not by her head. A dance on the steamer.

Thursday April 6
Quite pleasant with light winds from NW. Temp 8 AM 1°, 4 PM 13°. Captain Bernier was over awhile this evening. Mr Pelletier, the inspector of police, and Mr MacKean were over this afternoon.

Friday April 7
Very pleasant with light NW airs, hazy this morning. Temp 8 AM 5°, 4 PM 13°. We work on the boats making what repairs that are needed. I saw a small snow bird, and some partridges were seen today. Went over and spent the evening with Captain Bernier.

Saturday April 8
Overcast with wind increasing towards night and driving the snow. Temp 8 AM 18°, 4 PM 23°. Two of our natives went off today to hunt for deer and one sled returned having shot sixteen, bringing in seven of them, having left the remainder a short distance from here. They brought in seven salmon. A dance this evening here. Several of the saloon people spent the evening here. Opened a cask of bread. Opened a barrel of pork.

Sunday April 9
A stormy day, wind strong from north driving the snow. Temp 8 AM 15°, 4 PM 23°. The schooner came up quite a little, probably caused by the ice being carried down by the extra weight of snow which has drifted around us. Sent quite a large salmon to Mrs Moodie.

Monday April 10
Weather improving but still fresh breeze at times drifting the snow. Temp
8 AM 21°, 4 PM 27°. The sleds returned from Fort Churchill bringing the
mail for the police force only, no one else on the steamer getting any.
They arrived at the steamer at about 5 PM. They were thirty-three days
in going and remained nine days and were twenty-three days coming back.
Thickness of ice seventy-four.

Tuesday April 11
Mild but cloudy weather with light east airs. Temp 8 AM, 28° 4 PM 32°.
Major Moodie sent over a letter telling me what little news there was,
as only a few got letters. The mail had not arrived at Fort Churchill and
was not expected for two weeks after the sleds started to return, which
they were forced to do for fear of the rivers breaking up.

Wednesday April 12
Easterly winds light, and hazy. Temp 8 AM 29°, 4 PM 31°. A sled arrived
at the steamer from Chesterfield to trade. Spent the evening at the steamer
– they were having a dance. Got a few papers, some as late as the
fourteenth of December.

Thursday April 13
Wind is still to the eastward but moderate. Temp 8 AM 30°, 4 PM 20°.
Took a few pictures of native methods in anticooting.[7] Our sled returned
from deer hunting. They got two deer. I have rigged my boat with wire
backstays and think I shall like it.

Friday April 14
Wind moderate from NE. Temp 8 AM 27°, 4 PM 24°. Quite clear. Have
been at work making a pair of crutches for the carpenter. Captain Bernier
was over and spent the evening. He then sent over a bottle of wine to
drink the health of Canada's [Prime] Minister Sir Wilfrid Laurier.

Saturday April 15
Quite pleasant with moderate breeze from north, backing to NW. Temp
8 AM 14°, 4 PM 15°. One of our sleds with two natives and a man from

---

7 Few white men had the good fortune of witnessing an Eskimo anticoot or séance and
  fewer still, if any, were able to photograph the event. Here Comer appears to have
  persuaded a few of his natives to act out scenes from an anticoot. Two of his photographs of 'shamanistic performances of the Aivilik' – very likely taken at this time – were

our crew (Frank) have gone to Walrus Island to try for seal. A dance this evening.

Sunday April 16
Very pleasant with moderate breeze from north. Temp 8 AM 14°, 4 PM 11°. Most all of our natives have the itch, especially the children. Have been giving them sulphur and molasses. The natives have had to build new igloos, partly because the snow has settled the ice down [so] that the water covers the ice. Went over to the steamer this evening.

Monday April 17
Partly cloudy with fresh breeze north and NNW. Temp 8 AM − 5°, 4 PM 0°. Am now making a set of sails for my boat, having let Mr Ellis have my old one. The sled returned from Walrus Island with two seal. Mr Ellis and the cook with three natives have gone salmon fishing to the north of Cape Fullerton to stay a few days. Thickness of the ice seventy-four inches. The carpenter was able to sit up a short while.

Tuesday April 18
Very stormy wind, a gale from NW. Temp 8 AM 22°, 4 PM 24°. Working on the boat sails – the native women do the sewing.

Wednesday April 19
Still stormy with wind NE. Temp 8 AM 16°, 4 PM 20°. Captain Bernier was over and spent the evening. There was a dance on the steamer. Our carpenter was able to be taken up on deck for a short time and moved around on his crutches.

Thursday April 20
Weather improving though the wind is still NE and a little snow falling. Temp 8 AM 15°, 4 PM 23°. Still working on the boat sail. This weather is worse for the natives than the very coldest weather as then they can keep their lamps burning and keep warm, while now with this weather they are compelled to go without fire in order that the roof will not fall in upon them.

Friday April 21
Overcast. Moderate weather, wind NNE light. Temp 8 AM 15°, 4 PM 18°.

---

published later by Boas ('Second Report on the Eskimo of Baffin Land and Hudson Bay,' plates 1 and 2, facing p 511).

The water can be seen from part way up the rigging, the floe edge having broken off well in towards the reefs. Got off a cask of flour to be opened tomorrow.

Saturday April 22
Cloudy weather with light winds varying from SE to SW. Temp 8 AM 24°, 4 PM 28°. Finished the boat sail, which sets very well. Mr Ellis returned bringing fifteen salmon, most of them being small. Our natives got one seal at the floe. Started a new cask of flour. This being my birthday, the captain and saloon people of the steamer came over and brought [a] memorial gotten up by the artist Mr Mackean, which is very handsome. Mr. Vanasse also brought three volumes of a history of the nineteenth century as a present. This evening I spent on the steamer where Captain Bernier opened a bottle of wine. The carpenter was able to move back to his berth in the forecastle.

Sunday April 23
Very pleasant with light NW winds. Temp 8 AM 18°, 4 PM 24°. Took three plaster casts of natives' faces.

Monday April 24
Wind ENE increasing and snow falling. Temp 8 AM 4°, 4 PM 18°. Some of our natives have gone to get a boat which was left last fall part way towards Whale Point, also took a tierce of molasses to our camping place when we go to the floe later. We took the snow off from the deck and have been repairing the boats. Spent part of the evening on the steamer, where they are having a dance this evening. Had a very pleasant time. Thickness of ice seventy-five inches.

Tuesday April 25
Weather improving, wind NE to N. Temp 8 AM 6°, 4 PM 14°. Two more of our natives with their families arrived from Repulse Bay (nine people in all with seven dogs). They report game as scarce up there and the natives have eaten their dogs in some cases. The sled party which went after the boat returned this evening, leaving the boat at the floe edge. Stonewall and George are the names of the men who arrived with their families.

Wednesday April 26
Very pleasant with light north winds. Temp 8 AM 6°, 4 PM 16°. Put some paint on the boats. Our natives got one seal. Went over to the steamer this evening – they were having a dance.

Thursday April 27
Fresh winds from NE with some snow. Have been at work on casks and arranging them in the hold. We are getting things in readiness for the spring whaling.

Friday April 28
Very pleasant with light north and NE winds. We have overhauled our furs and repacked them. My intentions now are to use some very old casks to run the tryworks so that we may melt ice and get a good supply of water on hand.[8] Did some painting on the boats, also put up a new fly at the main truck. Our natives were not successful at the floe today.

Cask No. 1	35 musk-ox		Cask No. 5	27 musk-ox
Cask No. 2	23 musk-ox		Cask No. 6	20 musk-ox
Cask No. 3	31 bear		Cask No. 7	9 bear
Cask No. 4	164 fox, 61 wolves, 11 wolverine		Cask No. 8	25 musk-ox

Saturday April 29
Partly overcast with moderate NE winds. Dealt out clothing to the men. Working on the boats and the gear for whaling. A dance this evening here. Major Moodie was over and spent most of the evening.

Sunday April 30
Mostly overcast with moderate breeze from NNE, at night increasing. A little entertainment over at the steamer this evening on deck – quite a pleasant time.

Monday May 1
Blowing fresh from NE and drifting the snow. Weather improving. Thickness of the ice seventy-five inches. Got off some ice.

Tuesday May 2
Very pleasant with light breezes from NE to north. Two sleds of ours and a number of the native men have gone to hunt seal near Walrus Island

---

8 When ships were 'boiling,' or melting the oil out of whale blubber in the tryworks on deck, the blubber scraps from the try pots could be fed to the fires underneath. This fuel was not available when melting snow for drinking water, however. In the western Arctic driftwood brought down to the Beaufort Sea by the Mackenzie and other rivers was used for this purpose and for heating, but driftwood was scarce in Hudson Bay; hence Comer's decision to use old barrel staves.

to remain a few days. Took a couple of pictures of native mode of fishing through the ice.

Wednesday May 3
Very pleasant with light winds from WNW. The native men went out to the floe but there was no water to be seen. Coiled down the whale lines. A dance on the steamer this evening.

Thursday May 4
Very pleasant with light winds from NW to NE. Our natives returned with eight seal, seven from Walrus Island and one from near here. Had the sails loosed to dry. Opened a cask of bread. Major Moodie was over and took a picture of the steamer from the top of the house. Several of the women went off to ponds nearby fishing but got only one fish.

Friday May 5
Very pleasant with light breezes from NNE. Have been getting off ice to melt to fill up some casks with fresh water. Major and Mrs Moodie came over and took some pictures of our boats and natives, then came in and spent the evening.[9] Thirty years ago today I started on my first voyage whaling in the bark *Nile* of New London, Captain John O. Spicer.

Saturday May 6
Very pleasant with light airs westerly. Two boats were taken out to the floe, the natives returning. Got off what ice we have, which has fallen short this year, not having cut enough last fall. Major Moodie gave me some developing powder for bringing out the pictures on the glass plates. Took a couple of pictures of the boats when starting off, and this evening had good success in developing them.

Sunday May 7
Partly cloudy with moderate breeze from west. One of our native men is

9 The years 1903–5 witnessed an unusual intensity of photographic activity in north-western Hudson Bay. Comer, already a confirmed enthusiast, was more active than ever, employing flash powder for interior shots and acquiring tips from A.P. Low on developing technique. Low took a number of excellent photographs, some of which illustrate his book. Bernier possessed a camera. Mackean took snapshots at the beginning of the *Arctic*'s voyage and doubtless some in Hudson Bay as well, and some photographs were taken by Borden. Moodie also took photographs and during the winter of 1904–5 his wife produced very delicate portraits of Eskimos.

quite sick (Ben). The doctor is attending him. Had an entertainment in the steamer – the graphophone, and lantern slides shown by the major.

Monday May 8
Partly overcast with light westerly winds. Started the tryworks and melted ice and snow enough to make 120 barrels of water. Used wood for fuel. Nearly all of our natives went this morning to camp at the shore near the floe. One of the men (native) who is sick and Harry our head native, with their families, still remain here. Captain Bernier was over awhile this evening. This morning I got a couple of photographs of the natives' sleds starting off.

Tuesday May 9
Overcast with a moderate breeze from NE. Have been at work arranging casks in the hold. Went ashore this evening and called on Major and Mrs Moodie. Had a very pleasant evening. Opened a barrel of pork.

7

The Second Summer
(10 May – 8 September 1905)

Wednesday May 10
Very pleasant with the wind NE moderate. Mr Ellis and Mr Reynolds went with their boats to the floe this morning. Our natives came up with their dogs and helped them. They have taken seven oujougs or ground seal. Went over to the steamer this evening. Sold Mr Pelletier a bear skin for $15.00.

Thursday May 11
Cloudy with light winds from south, at night hauling to the SSW. We pumped salt water into all the casks which were empty, filling them partly full. A sled came up from the camp and got a few things which had been forgotten. Started in on three meals a day. Opened a cask of beef.

Friday May 12
Clear with a fresh breeze from NW. Today we hauled off a lot of small stone to put in the runs for ballast. Captain Bernier was over this evening.

Saturday May 13
Very pleasant with light NW winds. Stowed the stone ballast in the runs of the vessel. The sled came up and brought two barrels of blubber (seal). Took a picture of a family of the Kenepetu tribe – Peter and _____ [Susie?].

Sunday May 14
Overcast with the wind SW to south fresh. A sled came up from the floe reporting that one native (George) (Kan-ne-uke) had broken through the ice and lost his rifle. Have given him another. Spent the evening on the steamer. Took a picture of a family of the Kenepetu tribe – Charley and Mit-ka-u with their child.

Monday May 15
Overcast partly with wind moderate from SW. Today we unbent the squaresail and repaired it. Our natives are now all gone but the sick man (Ben) and his family. He is quite low but the doctor seems to think he will recover. Some sea gulls were seen today, the first this spring.

Tuesday May 16
Overcast with moderate breeze from south. We have taken down the house which we had over the after part of the vessel. Spent part of the evening on the steamer.

Wednesday May 17
Partly overcast with moderate NW winds. A little snow and rain last evening. Saw several sea gulls around today. Bent the mainsail today. Have sold to Major Moodie the lumber of the house for $50.00 – two thousand feet.[1] Ben our sick native was taken out to the camp near the floe so now all of our natives are out there. There are two families who have not yet come from Repulse Bay.

Thursday May 18
Overcast with moderate breeze from north to NE. Have been getting off some of the gear from the island. Major Moodie sent over and got the old lumber. Mr Moodie was over for a short while this afternoon and Captain Bernier was over this evening.

Friday May 19
Partly overcast with a little fog this afternoon, wind E to S. Major Moodie sent over the order for the money to pay for the lumber, $50.00 payable to the owners. Slushed the masts and cleared the snow from the vessel down to the ice. She is leaking now 900 strokes a day.

Saturday May 20
Overcast with light winds from the south. We left the schooner at 11 AM with our boat on a sled, having five dogs and a native driver, our carpenter having to remain at the schooner as his leg is not strong enough yet. Eider ducks are plentiful. Our boats have taken a number of seal and three

---

1 The 'house' was that built over the *Era*'s deck prior to wintering (see chap 2, n 1). Moodie used the *Era*'s old lumber to construct a 'commodious storehouse' approximately thirty by sixteen feet in area (J.D. Moodie, 'Report of Superintendent J.D. Moodie on Service in Hudson Bay ... 1904–5,' 10).

walrus. Two sleds arrived from Repulse Bay – Jimmy Palmer and Keckley with their families. Report deer plentiful at Yellow Bluff. Snowing at night.

Sunday May 21
Partly overcast with a moderate breeze from NW. Quite cold and raw. We have been cruising to the north while Mr Ellis and Mr Reynolds have gone to Walrus Island. As we have no timepiece in our boat it keeps us uncertain as to the time of day. The natives go to their tents ashore each night, the land being about one mile in.

Monday May 22
Overcast with a little snow. We cruised offshore. There is but very little ice to be seen. Opened a cask of bread which was brought down from the schooner. We have not in other seasons seen the sea so clear of ice as it is this. Ducks known as the south-southwest were first seen today for this season.

Tuesday May 23
Overcast with strong breeze and snow during the day. We did not go off. Mr Ellis and Mr Reynolds came up from Walrus Island. They had two seal and two walrus.

Wednesday May 24
Stormy wind north. Quite a lot of snow has fallen. At night hauled the boats up higher. The eider ducks are now beginning to mate.

Thursday May 25
Stormy day, wind north. Snowing most of the time. We can do nothing but eat and sleep.

Friday May 26
Weather clearing but blowing strong. Some snow.

Saturday May 27
Blowing very strong from north but at night weather clearing and moderating. Some snow drifting. The last of our natives to arrive from Repulse Bay came last night – Jack and his family. Now all our natives are here.

Sunday May 28
The weather has improved today. Mr Ellis and Mr Reynolds have gone to Walrus Island while I with a native boat have started for Whale Point.

Have had moderate breezes from NE to NW. Have stopped for the night about halfway to Whale Point, this small river being known to the natives as Kess-e-geer River.[2] Have taken three seal. At night pleasant. Not much ice to be seen offshore. Saw king eider ducks today, the first time this season.

Monday May 29
The day began quite pleasant with southerly winds. We went to Whale Point and found the ice well broke in close to the point. We then had dinner and started on our return trip. Stopped a little north of where we camped last night. Very little ice to be seen. Got one seal. Afternoon overcast, wind southwest.

Tuesday May 30
Overcast with light variable airs from the southward. We worked our way down to Cape Fullerton, returning with eight seal. We find that our sick native Ben is much worse and is not likely to live long. This is the man who pulled me out of the water when I had broken through the thin ice and to whose timely arrival I owe my life. This was in the winter of 1893–4.[3] I went ashore and saw him after supper.

Wednesday May 31
Stormy weather with snow, wind north. Have hauled the boats well up and in back of the reefs. A bad storm.

Thursday June 1
Weather moderating. Started and went to the schooner to prepare for the Southampton trip. Called on the major and his wife, also at the steamer. Settled with the doctor, whose bill for services and medicines was $78.68 for my crew and natives. Paid him in musk-ox skins (three). This is aside from what I have paid Sergeant Hayne for attendance on the carpenter, as he had charge of that case. In going to the schooner we went part way in the boats as the ice is breaking in fast this year. Came back at 5:30 PM. Weather coming in bad, wind SE with snow.

2 Probably the one now officially called the Borden River.
3 While sealing with the Eskimos Comer had gone through thin ice up to his waist. He tried to extricate himself but only succeeded in losing his rifle. Ben hastened to Comer's assistance, managed to get him out, put him on a dog sled and drove as fast as possible to the ship, about four miles away. The air temperature was between − 29 and − 35°F at the time (George Comer, Manuscript Journal on Board the *Canton* 1893–4, 23 February 1894, George Comer Papers, East Haddam, Conn).

Friday June 2
Pleasant though blowing fresh from NW. Our native Ben who has been sick died this afternoon. My boat's crew and I went up to the tent as he was a man whom all liked. I helped carry him away and assisted in the burial. His two wives, brother Harry, and his two children, also Gilbert and Sam, we built stones around and over him. Our mittens had to be thrown away. The ropes which held the deer skin around him had to be all cut so that he could rest easy.[4]

Saturday June 3
Fresh breezes with snow squalls. The two boats returned from Walrus Island, have taken one walrus, two seals and five deer. On account of Ben's death all natives must stop work for three days.[5]

Sunday June 4
Fresh breeze from NW, a little snow. Most of the men went up to the schooner to change clothes. They report one of the steamer's natives having died on the second (Old Charley).

Monday June 5
Pleasant this forenoon. At night overcast with variable winds from the southward. Went to the schooner and got provision for our trip to Southampton.

Tuesday June 6
A very stormy day, wind backing from NE to north. The ice breaking up from the effects of the swell, we had to move and found it difficult to find a place where we could haul out with safety. Much snow has fallen.

Wednesday June 7 Whale Point
Moderate breezes from NW to N but quite cold. Overcast most of the time. We left Cape Fullerton at 6:45 AM – four boats – leaving Harry and Sam to bring along the women and children, we being bound for Southampton.

4 Cutting the thongs around the deerskin-covered body enabled the soul to leave the body (Franz Boas, 'Second Report on the Eskimo of Baffin Land and Hudson Bay,' 486).
5 A number of other rituals and taboos were connected with death. The customs varied from one group to another. The Cumberland Sound Eskimos, for example, for three days following a death could not wash their faces or hands, cut hair, trim fingernails, or feed dogs (ibid.).

We arrived at Whale Point at 11:15 AM, a quick run. We shall start from here to go across in the morning, weather permitting. In coming up today the water froze on the boats quite a little. Took snapshots of the boats laying at the edge of the ice and when hauled out putting the covers on.

Thursday June 8
A stormy day with snow driving from the NE. At night clearing with the wind backing to NW. We are quite uncomfortable with the cold. Had to haul the boats up higher.

Friday June 9 Whale Point
The day set in with a moderate breeze and snow squalls from NW. We started for Southampton at 7:30. When about twelve miles off we raised a whale but only saw it two risings when we lost the run of it, but worked back toward Whale Point thinking that the whale might work in that way or toward Yellow Bluff. Mr Ellis and Reynolds went up there while Gilbert and I stayed near Whale Point. At 4 PM a thick snow storm from south – had to haul our boats up on the land.

Saturday June 10
Blowing strong from the NW, the wind hauling during the night. It was the coldest night we have had. Most every night the coffee kettle freezes up but last night more than ever. This morning Mr Ellis came back towing Mr Reynolds' boat. The ice in breaking up where they were set them adrift and Mr Reynolds got his boat badly stove up. They were out nearly all night and all their men had ice on their clothing, the boats and gear being covered. We gave them our dry clothing and hot coffee and they got into dry sleeping bags at once. The boat is in very bad state.

Sunday June 11
Overcast most of the time with fresh breeze from NW. A little snow and quite cold. Have been at work on Mr Reynolds' boat. We cannot make a good job of it with the things we have to work with and it is so badly damaged. Cannot work without mittens on. The men from Mr Ellis and Reynolds' boat have to keep in their sleeping bags till their clothes get dry.

Monday June 12
We left Whale Point at 8 AM after finishing the work on Mr Reynolds' boat. Started again for Southampton. Arrived there at 6 PM. Light winds

and calms. Could not get near the shore ice on account of so much broken ice on the flats, which make well off. Had to haul out on small pieces at high tide and at low water no water to be seen. Snowing at night, wind NW increasing.

Tuesday June 13 Cape Kendall
This morning we worked our way out of the ice to the edge where we stopped on account of so much wind and sea, but at 1 PM shoved off and worked to the south. At about 3 PM Mr Reynolds raised a whale and in the next rising struck it. We went in second boat but the whale knocked us to pieces in short order, Brass the boatsteerer being knocked overboard before he could throw the second iron, then hitting the boat twice again and on the fourth sweep of the flukes I was thrown well up in the air and off one side, then the whale started off spouting thick blood. Brass was picked up by another boat and I managed to get back to my own, where I was pulled in by my men. The boat's bottom being all open she filled at once. The native boat came and helped us. I lost a fine pair of glasses given to me by Commander Low last winter (when he was here), also a pocket barometer, five oars and a number of small articles as well as all the provision. Mr Ellis and Reynolds got the whale killed and towed it into shallow water. We hauled our boats out on a cake of ice and fixed the bottom best we could. Changed my clothes. We got supper later, then towed the whale in on the high tide at 11 PM, leaving it about a mile from the shore ice. Anchored, then came in and hauled out on the shore ice at 1 AM.

Wednesday June 14 Cape Kendall
Went off to the whale at 1 PM with three boats, leaving mine for repairs. We got the whalebone all cut out by 6 PM then I came ashore, leaving the boats to come in with the bone when the tide arose high enough, which was at 11 PM. It was a bull whale, bone being $9^{1}/_{2}$ feet long [see appendix G]. We are hauled out on the ground ice, which extends from the shore some three miles off. Wind NW moderate, a little snow. Wet clothing and sleeping bags.

Thursday June 15
Spent the forenoon in fixing up my boat then when the tide got high enough we pushed off and started to work back to Cape Fullerton by the way of Whale Point. When about ten miles off raised a whale going south. Too

fast for us, though we worked after it some two hours. We then kept working towards Whale Point but by 9 PM the wind came out fresh from NW and the weather not looking good we turned back and reached Cape Kendall at 1 AM, where we managed to get to the shore ice close to the land.

Friday June 16 Whale Point
We left Cape Kendall at 1 PM at high tide and had light airs from SW with calms. Arrived at Whale Point at 2 AM. Found our natives had taken a small whale. Sam struck first. Took the whale this forenoon.

Saturday June 17 Whale Point
The natives cut the bone out of their whale on this morning's low tide. We have split it all (both heads) and stored it in the house till we come up with the schooner. The wind is strong from south. The natives have caught a number of seal and salmon from the ponds.

Sunday June 18
Blowing very strong from SW so that we cannot go to Cape Fullerton. We are now getting rested after our trip. We keep a lookout for whale from the house.

Monday June 19 At the schooner
The wind moderated and followed by a calm, then breezed up from NE. We left Whale Point at 1 PM and came to the schooner at midnight, leaving the boats on Beacon Island. We got nine eider ducks' eggs coming down. Sam's boat came with me.

Tuesday June 20
We have been getting provision ready. Have brought my boat to the schooner in order to do better work. Captain Bernier has been over.

Wednesday June 21
Cold and raw NNE winds, some rain. We have been at work on the boat. Major Moodie came over awhile.

Thursday June 22
The weather has improved so that we could work on the boat to some advantage. Nearly finished. Have put in two new planks. Went over to

the steamer this evening. Cut through the ice and got an anchor down.

Friday June 23
A very pleasant day with light airs from s to sw. We finished the boat and at 6 PM took her to the water and loaded her up, starting for Whale Point. We had calm all night. Stopped and went to Ben's (the native) grave, then stopped at the egg island and got about 100 eider ducks' eggs. Arrived at Whale Point at 4 AM the twenty-fourth. Got an oujoug (large seal) in the ice.

Saturday June 24
We arrived at Whale Point at 11 AM. Had a light breeze from the sw. Got an oujoug. Divided up the provision with the other boats. We _____ _____ [feel tired?] out. Mr Reynold's crew – they saw a whale when they were cruising.

Sunday June 25 Cape Kendall
We left Whale Point at 8 AM and had a good run across. When we got to the place where we had taken the [whale] we separated, Mr Ellis and Mr Reynolds going to the north while Gilbert (native) and myself worked to the eastward, bound to the Scotch station.[6] Hauled out ten miles east of Cape Kendall on the ice on the east side of the long reef. Partly clear, light sw to w breeze. Shot a bear. Harry and Sam are to work up to Wager River.

Monday June 26 Near Scotch Station, Southampton
We left Cape Kendall flats at 8 AM with a moderate breeze all day from south to west. Arrived near the Scotch station at 8 AM. We are about fifteen miles south of Manico Point and about ten miles north of the house.

Tuesday June 27
Pleasant but with quite a swell heaving in on the beach so that we have not put off. The natives went hunting, and got three deer. Gathered a few native implements from native graves.

Wednesday June 28
We got off at 8 AM and got out through the surf, which was more moderate

6 This was the whaling and trading station run by the Scots between 1899 and 1903. It was situated about 25 miles south of Manico Point at Cape Low (see chap 1, n 21).

today. Pulled down to the Scotch station. Of course no one lives here now, the place being abandoned in 1903. Hauled out. We keep a lookout from a high frame tower.

Thursday June 29
There has been a very strong breeze from NE all day so that we could not go off, but have kept a lookout from the tower. We would like it better if there was ice around, then it would remain smooth in bad weather.

Friday June 30
Two years out today. Strong winds from SE afternoon. Got one loon's egg, two gulls' eggs and several small eggs. Two king eider duck eggs were got.

Saturday July 1 Fifteen miles north of station
Some rain with fog and fresh breezes from SSE. We left at 10:30 and worked back to the northward. We hauled out at 1 PM, still foggy with some rain. Got six swans' eggs, two loons' eggs, some native implements. There are a number of old bone and stone houses here.

Sunday July 2
We started at 9:30 and came across the Bay of Gods Mercy. Had moderate breeze from west. On this shore the ice extends off about two miles. Took some pictures of native house before leaving this morning.

Monday July 3 North side of Bay of Gods Mercy
Strong wind from south with much rain during the night and forenoon. There was quite a heavy pack of ice crowding up at high tide. We pushed off and pulled along the lead as fast as the ice set off. Came along about five miles then hauled out near the land. The natives have gone off to look for deer. Got three swan eggs.

Tuesday July 4 South shore of Cape Kendall
Pushed off at 3:30 AM and came along about five miles and then got breakfast and later when the tide turned we came past the lowland which seems to extend across the land to the NE of Cape Kendall and stopped at 4 PM but could not reach the land, the tide having fallen. We hauled out on a grounded piece of ice. Have had moderate SW winds. The natives got three deer last night. Partly cloudy.

Wednesday July 5 Cape Kendall
We got away at 1:30 AM. Had a moderate breeze from SE. Had to go offshore about twelve miles to get clear of a long reef which extends offshore. Arrived at Cape Kendall. At 3:30 PM we saw two bears at rest on a snow bank. We stopped and got them after quite a chase. Mosquitoes made their first appearance.

Thursday July 6 North of Cape Kendall
We left Cape Kendall at 4 AM. Had light wind at first then later it became fresh from NE. We came about thirty miles following the land. At night breeze more moderate.

Friday July 7
We pushed off again at 5:30 AM but had to haul out at 7:30 on account of a strong head wind from NE. We are near where there are a number of old sod and bone houses. Got a number of articles of native make for the museum in New York. Our natives got eight deer.

Saturday July 8
The wind being fresh we did not start till 5 PM when the weather moderated. We came up about six miles then hauled out, the tide having fallen some. Light head winds NW. This shore running NNW from the bend ENE of Cape Kendall.

Sunday July 9 Nuvuk [Point]
We pushed off from Southampton with the wind moderate from NE. Later it worked to the NW. We worked up the coast till night then started across, where we arrived at midnight. We could not haul out but anchored and slept till high tide. Light NW winds in coming across.

Monday July 10 Near Wager River
We have worked from Nuvuk up towards Wager River and at night hauled out at a place where we once took a whale (in 1899). Got a number of eggs. Moderate NE winds.

Tuesday July 11
We pushed off at 8:30 AM and started for Whale Point where we arrived at 7:30. Had moderate to fresh NE wind. Found all four boats here, nothing seen. It is reported that the steamer sailed the fifth and the same afternoon one of the police force which were left at Cape Fullerton was drowned.

Wednesday July 12 At the schooner
We left Whale Point with a light breeze and arrived at the schooner at
10 PM (six boats). Found the schooner leaking very much.

Thursday July 13
Light southerly winds, very pleasant. We have endeavored to get at the
leak by getting the schooner down by the head. Have hung three casks
of water from the bowsprit and carried the stone from the after run forward.[7] This does not help.

Friday July 14
We have done what we could to locate the leak but cannot find anything
that looks satisfactory. We have put the stone back in the run and brought
off all the heavy stores. Have painted four of the boats and repaired Mr
Reynolds' boat. Southeast winds with a little rain. The schooner has been
painted inside while we have been away.

Saturday July 15
We have finished loading what we have to take on board and are now
quite ready to leave for Repulse Bay. The weather today has been stormy
– SE to east winds. Sergeant Hayne is in charge of the camp here. He
was on board this evening. He has now two men with him, one man
having drowned the day the steamer left – Joseph Russell. They were (he
and another man) out after eggs and were using a small boat in going
across a pond near the house.

Sunday July 16
Blowing quite strong during the night and forenoon from ENE. Afternoon
moderating, wind still the same. We towed out of the inner harbor at 5
PM and we are now in the outer harbor. Light airs and pleasant.

Monday July 17
Light airs and calms during the day. Our natives left here this morning
to return to Whale Point (three boats). We have finished scraping the
masts and also tarred down the rigging. Today light airs from SE to SW.
Mosquitoes have been quite thick around the vessel.

7 The stone was presumably ballast.

Tuesday July 18
Light airs and calms till 4 PM when we got under way with a light air from
SW. Sergeant Hayne and one of his men (Stoddard) are going as passengers
with us to Repulse Bay. Have been painting the mast heads and ends of
booms and gaffs. We have three native boys and one woman on board.
We are to stop at Whale Point and take on a number of natives with their
dogs and belongings while the three native boats will go along the shore,
the Sergeant's boat keeping company with them.

Wednesday July 19
Had moderate breezes through the night from SW and arrived at Whale
Point early this morning. Sent the three boats in to the natives but it was
noon before we got everything off, having to make two trips each boat,
the other three boats having proceeded to Repulse Bay. We have a number
of natives with dogs, sleds and tents. We are now on our way up the
Welcome with a moderate breeze from south. Took our bone on board
and weighed it, making 1,697 pounds. The two whales: the small one 241

$$\begin{array}{r} \text{the large one } \underline{1456} \\ 1697 \end{array}$$

Thursday July 20
The wind died out towards morning, hauling to the west, and today
worked to the NE so that we have gained but little. We are about twenty
miles north of Wager River. Pleasant weather. Moderate swell coming
up from the southward.

Friday July 21
Very light airs from north to NE and at night a light air from west. We
are now near Beach Point. There is some bay ice though not enough to
hinder so far. Got a walrus. Opened a cask of bread.

Saturday July 22
Last night we got becalmed in the ice which closed around us and we
could not help ourselves. We took in all sail after working and running
lines till 2 AM. The ice on the rising tide carried us close into Beach Point
at 4 AM and at 5 AM we passed over a reef which at the time seemed as
if it would wreck us. We finally by towing came to open water and anchored in nine fathoms of water where we lay till noon, then got underway
with a light air from south and worked our way over to the Southampton
shore. We have a light breeze from south-southwest and are now working

again into Repulse Bay. The ice has only recently broken up and the whole bay seems to be full. Light, fine weather.

Sunday July 23
We kept in open water during the night near the Southampton shore and this morning started to work up through the ice. Had light airs and calms during the day with the wind hauling to the westward. We arrived at the harbor at 10 PM and anchored. Had to come to our anchor on account of the wind being ahead and the tide running out. We then run a line to the Scotch vessel (*Ernest William*) and hauled up near her and anchored in eight fathoms. Our boats arrived here yesterday all well. It is reported that two whales have been seen here. One of our natives fell overboard, but was pulled into a boat laying alongside (John L.) Very warm weather for this country. We have had a number of very fine days and all of them quite warm.

Monday July 24
Partly cloudy with light variable airs south to NW. We are preparing to start to Lyon Inlet and also painted the schooner outside. Boiled out about twelve barrels of blubber. Called on the master of the smack, Mr Forman.[8] Sergeant Hayne has taken up his quarters ashore, in a tent.

Tuesday July 25
Pleasant most of the time, wind varying in force from NW. We ran down the oil and have got quite packed up for a start to Lyon Inlet. Some of the Netchilic natives have brought a very little trade.

Wednesday July 26 Frozen Strait
We left the schooner, three boats – Mr Reynolds, Gilbert (native), and myself. The other three boats are to leave tomorrow. We are bound to Lyon Inlet. Have had moderate NW and west breezes. Came over to the Blue Lands and down to Frozen Strait. We are hauled out on what is

8 William Forman came out from Scotland on the steamer *Active* and with three other men joined the *Ernest William* on 10 September 1904 at Repulse Bay to man the vessel during the winter. Forman appears to have been in charge. In 1905 the ketch (or 'smack') was towed to Lyon Inlet by the *Active* to winter (see journal entry 26 August), this time under the command of J.W. Murray (brother of Alexander Murray, master of the *Active*). Forman returned to Scotland on the *Active* (*Active*. Abstract logbook 1904 and 'Agreement and Account of Crew' 1905. General Register and Record Office of Shipping and Seamen, Hayes, London).

called Bushnan Island, though there are several. Are on the second one from the north.

Thursday July 27
Pushed off at 7 AM (high tide). Met a party of the Scotch natives who were camped on the next island. We sailed up through Gore Bay and over towards Lyon Inlet where we stopped on a round island known to the natives as Im-may-u-er-tuck. Wind SE, no ice to speak of.

Friday July 28
We left the island (Im-ma-wet-luck) at 7 AM. Had light SE winds which later increased to a strong breeze. We reached Winter Island[9] at 1 PM and at 4 PM came over to Cape Edwards (Lyon Inlet), where we met the two boats' crews of the Scotch smack. They report having struck a whale on the eleventh but lost it.

Saturday July 29
A very stormy day with hail and snow. Wind NE to north.

Sunday July 30 Bull whale 612 pounds bone
A very stormy night and forenoon. This afternoon more moderate. Raised a whale at 3 PM and pushed off though blowing strong, but weather improving. Got fast at 8:30 PM (Gilbert). The whale, evidently mistaking the boat for a piece of ice, came up under it. Got it to the shore at 10 PM high tide. Wind backed to south, stormy appearance.

Monday July 31 Lyon Inlet
A very heavy gale from south with rain. As the tide got low this afternoon we cut the bone out – six feet ten inches. The Scotch natives are working with ours saving the meat and blubber. At night weather clearing a little.

Tuesday August 1
Some fog and rain. We were scraping bone till 3 PM when a whale was raised. Pushed off and saw two. Mr Reynolds got fast to the smaller one.

9 Winter Island was where Parry's ships *Fury* and *Hecla* spent the first winter during the Northwest Passage expedition of 1821–3. Many names in the region, including Bushnan Island, Vansittart Island, Gore Bay, Lyon Inlet, Farhill Point, Safety Cove, and Cape Martineau, were established by the expedition (William Edward Parry, *Journal of a Second Voyage for the Discovery of a North-West Passage … in His Majesty's Ships Fury and Hecla …* [London: John Murray 1824]).

Got it ashore at 6 PM low tide. The whale was raised by a native boy living here. Later he got his box of tobacco – this is a prize given to the person who first sees a whale and it is caught. Length of bone six feet ten inches.

Wednesday August 2 Lyon Inlet
This morning a whale was seen at 1 AM. Pushed off at 1:30 AM. Had light airs with mist. We lost run of it and came back and cut the bone out of yesterday's whale. The weather has become very stormy, NE with rain and snow, fog and hail. Blowing a gale.

Thursday August 3
Raised a whale at 2:30 AM and pushed off at once. I went on to one but he settled before reaching. The boatsteerer darted but did not get fast – pricked the whale a little. We have seen four whales. It is reported that the Scotch people got two whales. Some fog and rain during the day. Our other three boats came today, this evening. Wind SE varying in force. A little ice drifting in.

Friday August 4 Lyon Inlet
Raised a whale at 2:30 AM and pushed off, some fog and rain at the time. Mr Ellis got fast and after having killed it and while towing it ashore raised another, which I went after and got fast to. Got it ashore at 3 PM, then we got our breakfast. Mr Ellis, Harry and Sam are taking care of their whale while Mr Reynolds, Gilbert, and my boat's crew are taking care of our whale. Wind ENE fresh. The bone measured eight feet ten inches of Mr Ellis' whale and nine feet of my whale.

Saturday August 5
Blowing strong from NE. Finished cutting out the bone and have been splitting and scraping the gums off and tied it in bundles. Weather improving so that we can dry our clothes.

Sunday August 6
Blowing a strong gale from NE but quite clear. The natives have saved all the meat and blubber of the four whales. They expect to camp here this coming winter as they are now sure of dog food and oil.[10] Brought all the bone together – six bundles.

---

10 A whale carcass with only the baleen removed could provide subsistence for Eskimos

Monday August 7
Wind still strong from NE but moderating. There is a pack of ice working in so deemed it best to take our bone to the schooner. We loaded up at 4 PM and came over towards Gore Bay to Im-ma-we-tuck Island. Hauled out at 9 PM. Did not have any great trouble in getting through the ice.

Tuesday August 8
Got started at 4 AM. Have had fog during the forenoon then pleasant with light airs. At 6 PM, when we hauled out at the salmon stream on the Blue Lands, found some natives here. They have taken a large number of salmon so we have all we want to eat for supper.

Wednesday August 9
We arrived at the schooner at noon. Light variable winds. The bone weighed 3,764 pounds. Stowed it in after run. The Scotch boats arrived this evening at the smack. The steamer has not arrived here as yet. We are leaking about 5,000 strokes a day.

Thursday August 10
We left the schooner, Mr Ellis, Mr Reynolds, and myself – three boats – to return to Lyon Inlet. The three native boats are to come on tomorrow as soon as provision can be cooked. Have had light airs and calms. Came as far as the SE cape and hauled out at 8 PM.

Friday August 11
Started at 7 AM. Have had light variable winds. Arrived in Gore Bay at 9 PM. Had some fog during the forenoon, very little ice. We stopped at a Scotch native camp – John Bull – as we came along.

Saturday August 12
Started from Gore Bay at 7 AM and came to Cape Edwards, Lyon Inlet, at 9 PM. Shot one bear. Light airs and calms. The natives here have seen three whales. We heard one this evening.

---

for a long time. There was meat for humans and sled dogs and the blubber contained several tons of oil which could be used for heat, light, and cooking. Whale ribs and other bones could be used for building houses. In addition, the rotting carcass would probably attract scavenging polar bears, arctic foxes, and birds whose meat, skins, and fur could be utilized. There are several recorded instances of Eskimo families or bands moving to live close to whale carcasses, one way in which the whaling industry influenced the geographical distribution of the native population.

Sunday August 13
Raised a whale at 7 AM pushed off and soon Mr Reynolds got fast, then
Mr Ellis. We got the whale ashore at 11:30 AM, the tide having fallen a
little. This afternoon cut the bone out – nine foot three inches. Our other
three boats came this afternoon, also three Scotch boats. Moderate NW
breeze. Four other whales have been reported as seen today.

Monday August 14
Raised a whale close to the shore. All the boats put off but in the rush
some one of them must have scared it, as it was not seen again. There
are five Scotch boats here and with our six make quite a show. A shark
was taken today. It measured nine feet five inches in length, the first ever
seen here by the natives.[11]

Tuesday August 15
We came across to the west shore, Cape Martineau – Mr Reynolds and
I – while Mr Ellis and Gilbert have gone to Winter Island, Harry and Sam
stopping at Cape Edwards and finishing the bone scraping. Pleasant,
moderate NW breeze.

Wednesday August 16
We left Cape Martineau and came to the first island south of it, about
eight miles, where we saw two whales but having light airs and calms
could not get near them. At night saw five at one time. They are close in
to the shore. There is a gull rookery here at this point. A little ice. Light
airs SE, pleasant.

Thursday August 17
We pushed off at 2 AM. Had light airs and calms. There were at least six
whales around but we could do nothing with sails. We then took them in
and downed masts, then by laying near a point of land and waiting we
got a chance at one which came up near my boat. By sculling with the
steering oar we got onto it and fastened. Got my hands badly burned

11 The Greenland shark (*Somniosus microcephalus*) is a large but passive fish which is
   hunted in Greenland for its skin, liver, and meat. It is commonly found in Davis Strait
   and Baffin Bay, and individuals have been observed in Hudson Strait and Ungava
   Bay from time to time (M.J. Dunbar, 'On the Fishery Potential of the Sea Waters of
   the Canadian North,' *Arctic*, vol 23, no 3 [1970]: 163–5). This occurrence, however,
   was far beyond the normal range of the species.

holding the line. We pulled the whale up dead – small cow whale. While waiting to get the chance our native was ashore skinning a bear he had shot. Length of bone six feet three. Whale feed is very thick and we can hear a whale blowing now at 9 PM.[12] We are laying off anchored as the tide is low and too hard to pull the boats up. We put the covers on and sleep.

Friday August 18
Have had a moderate breeze from west but have not seen a whale today. Took the bone out of the one we had. Took the bone and bear skin to the passageway leading from Lyon Inlet to Gore Bay and left it on a small island so as to take it on our return trip. We are now back on the second island. Pleasant weather.

Saturday August 19
We returned to Cape Martineau where we found Mr Ellis and Gilbert. Have sent Gilbert back to tell Harry and Sam to take the bone and bear skins to the schooner and get more provision. Mr Ellis, Mr Reynolds and myself are going to cruise around these islands. The men saw a whale and we set out after it but did not see it again. Light airs from west.

Sunday August 20 Vansittart Island
We came over to this island this forenoon. The place where we are now is at a very narrow part of the island. A boat could be taken across if needed. The wind being NE and strong we have stopped here to try and get some deer. Jack got one this afternoon which is quite a treat to us. Mr Ellis has the scurvy.

Monday August 21
Still blowing strong from NE though pleasant otherwise. I took quite a long walk. There are indications that in the long past many natives lived here and at this narrow part of the island used to shoot the deer by keeping back of stone blinds which still remain.[13] We came to the narrow part of

---

12 When feeding, the Greenland or bowhead whale fills its mouth with water and then expels it, catching the zooplanktonic krill or 'whale feed' on its 700-odd baleen plates. The several species of euphausid crustaceans and pteropod mollusks that comprise the krill often occur in visible swarms near the surface, and the whaling captains took this to be a promising sign that whales were present in the area.

13 A common Eskimo method of hunting caribou was to erect two converging rows of

the island this afternoon at high tide with the boats. Got a number of old native implements made of stone.

Tuesday August 22 Frozen Strait
Weather quite moderate. We pushed off at 2 AM and raised a whale soon after. It was going to the north quick. Soon lost run of it. We came around the south end of Vansittart Island. Have taken five bears this afternoon. Got one deer this forenoon. The young bear skins are quite good now. Light airs NE, then variable.

Wednesday August 23
Pushed off at 4 AM, light airs and calms. Got six bears and might have taken more had we cared to stop. We wish to get back to Gore Bay. A little rain. We have stopped abreast of the Duke of York Bay.

Thursday August 24
Started at 4 AM and came up around the north end of the island into Hurd Channel. Light airs and calms. Have hauled out on the north shore Hurd Channel near turn to go to Gore Bay. High tide 5:30 PM.

Friday August 25
Started at 4 AM and came through Gore Bay and arrived here at 3 PM where we caught our last whale. Five Scotch boats came here this afternoon. They have taken one very small whale a few days ago. Our three native boats arrived here this evening from the schooner. They report the Scotch steamer *Active* having arrived at Repulse Bay on the morning of the twenty-fourth. They (our boats) did not bring any provision or mail.

Saturday August 26
Bad weather. Strong wind from SW, raining. This afternoon the steamer *Active* came to Lyon Inlet towing the ketch.[14] She has anchored in one of the bays on the west shore for the night.

---

stone cairns (inuksuk) resembling people, through which the caribou would move or be driven towards an ambush on land or in the water.

14 The object was to establish the *Ernest William* as a base for whaling operations in Lyon Inlet, one of the last resorts of the bowhead whale in the Hudson Bay region. The steamer towed her over because narrow passages and swift currents between Frozen Strait and Lyon Inlet made the voyage hazardous for a vessel without power. The *Ernest William* wintered in Lyon Inlet in 1905–6 and probably again in 1906–7. She sailed back to Scotland in 1910, having spent eight consecutive years in Hudson Bay and adjacent waters.

Sunday August 27
Quite a little rain fell last night. Foggy this morning. Mr Ellis and Mr Reynolds have started to the vessel to get provision and mail. This afternoon Harry and Sam have gone to Vansittart Island to look for deer. Could see the steamer towing the ketch up towards Safety Cove this morning.

Monday August 28
Blowing strong this morning but later moderate, wind north. Hauled the boats up higher as there was quite a surf running. Our food is getting quite low but hope our natives will bring some deer meat back from Vansittart Island.

Tuesday August 29
Moderate breezes from west to NW. Threatening appearance of the weather but still moderate. Harry and Sam returned with eight deer and one bear.

Wednesday August 30
Wind generally fresh throughout the day. Could see the steamer towing the ketch into a bay north of Cape Edwards. We are having plenty of deer meat just now.

Thursday August 31
As we have not seen anything here we came back to Hurd Channel, thinking to meet Mr Ellis and Reynolds and then go over to the Duke of York Bay if weather permits. Wind south to west. Hauled out at 7 PM [at] Hurd Channel narrows.

Friday September 1 Bushnan Island, Frozen Strait
Mr Ellis and Reynolds met us this morning in Hurd Channel then we all pulled up to Bushnan Island (middle one). Let the natives return to the vessel while we stop here to go over to the Duke of York Bay. Got letters from home, also potatoes and onions.

Saturday September 2 Southeast cape of Repulse Bay
This morning we started to cross over Frozen Strait and got as far as Passage Island when the wind came out SW very strong so we put back and, as the tides run very strong here, we could do nothing. Looks now as though we should have to close up the voyage.

Sunday September 3 At the schooner
Arrived at the schooner at 4 PM. Light winds. Coming on a rain storm
from the SE, wind increasing.

Monday September 4
Have been coopering casks to fill with fresh water to raft off. Have been
paying off the natives. Fresh gale from SE.

Tuesday September 5
Blowing fresh from NW, moderate at night. Have got off a raft of water,
also filling up salt water casks.[15] Gave the natives a cask of bread and
three tierces of molasses.

Wednesday September 6
Finished getting off the fresh water and have given the natives what we
had to spare so that we are now all ready to leave for home. Today wind
fresh from SE, raining this afternoon.

Thursday September 7 Repulse Bay
A very stormy day. Wind heavy from NE to north. Fog, rain and much
snow – the land is well covered. Let go the second anchor. Weather getting
colder. Scuttlebutt frozen up. The native small boat went adrift. We are
all ready to leave but cannot see but a little ways.

Friday September 8
The weather moderated the latter part of the night. This morning we got
ready by pounding the ice off the rigging and clearing the deck of snow.
We then got underway at 7 AM, wind NW moderate but squally. The
natives gathered on a point of land and we gave them three cheers and
they returned us the same as we passed them. The winds have been light
and have backed around to south, then east, then north, and then SW.
Thick snow squalls at times but weather improving as the moon came
up. We are about ten miles south of Beach Point.

15 Salt water may have been used as ballast, being easier to load than rocks.

8

Voyage Home
(9 September – 15 October 1905)

Saturday September 9
Had moderate gale during the night. Have worked down under two-reefed mainsail, wearing each time we went around. Wind SW with heavy snow squalls. At 6 PM passed Wager River. At noon our latitude was 65° 35′N. The weather has improved. Shook reefs out of mainsail this afternoon.

Sunday September 10
The weather has been more moderate. Wind SW, partly clear. We are down to Yellow Bluff, have made but little distance this afternoon. It is my endeavour to stop at Cape Fullerton and take what mail the police may wish to send and also leave a boat there for a native who has partly paid for it.

Monday September 11
Weather more moderate. Wind SW with thick snow squalls at times. At night the wind has let go. We are near Cape Fullerton. A moderate swell from south which is setting us in towards the land. We have eighteen fathoms. Sky mostly overcast. The schooner is leaking considerably.

Tuesday September 12
The wind died out and then sprung up from NE last night. This morning on standing in towards the land found that we were south of Cape Fullerton and also to the west, the wind being strong from NE. I found that after making a couple of tacks that it would use up all the day to get there. We have a boat which belongs to one of the natives (name Smiley) who has partly and nearly paid for it. I wished to leave it for him but could not.

I also wished to take any mail that the police might wish to send. We kept off at 10 AM, came over to Cape Fullerton and are now running for Cape Southampton, Coats Island. Heavy snow squalls this morning, later weather improving. We are now bound home and take our departure from Cape Fullerton. Sergeant Hayne had given me a package of mail to forward in case I did not stop, but I am disappointed about the native not getting his boat.[1]

Wednesday September 13
Have had strong winds with heavy snow squalls during the night and today from NW, irregular swell. Made Cape Southampton at 3 PM, passed Carys Swan Nest at 5 PM. Wind fresh NW. Two reefs in mainsail. Generally overcast.

Thursday September 14
Had very strong breezes off the land (Coats Island). Had to put one reef in foresail and tie up the flying jib. Mansel Island could be seen from

1 The events following the *Era's* failure to deliver the whaleboat to Smiley illustrate Comer's respect for oral contracts made with the Eskimos for employment or trade. In 1906, prior to embarking on his next voyage to Hudson Bay, Comer wrote to the comptroller of the Royal North-West Mounted Police, explaining that the *Era* had been sold by Thomas Luce to F.J. Monjo, and that the Luce firm had refused to assume responsibility for the debt to Smiley (Comer to White, 2 July 1906, RCMP Records, vol 321, file 566). Comer felt that as the owners had not hesitated to take the furs paid by Smiley they had an obligation to supply the boat. But Comptroller White insisted that the responsibility was Comer's because he had arranged the transaction and had then 'failed to give the compensation agreed upon' (White to Comer, 20 July 1906, ibid.). If Comer did not complete his part of the bargain, he warned, 'the native of course will have the usual recourse through process of law' (White to Comer, 12 July 1906, ibid.). Comer recorded all the facts before a public notary and persuaded his new employer, F.J. Monjo, to write to White to see what could be done to induce Luce to settle the matter. White would do nothing.
    Comer had vowed that Smiley would get his boat even if he had to pay for it himself, and in the end he purchased a whaleboat (probably costing about $160) and loaded it onto the *Era* for his next trip to Hudson Bay, along with two boats to be delivered to the RNWMP. Unfortunately the *Era* was wrecked off Newfoundland (see Epilogue). The boats were swept ashore, severely damaged. One of them – possibly the one intended for Smiley – was purchased from the Canadian government by a salvor for ten dollars less five dollars 'expenses' (Woodhouse to White, 1 September 1906, ibid., vol 311, file 130). In 1907 Comer headed for Hudson Bay again with another boat on deck, and finally succeeded in delivering to Smiley one whaleboat 'for a debt contracted by the master of the schooner *Era* for Thomas Luce & Co.' (*A.T. Gifford*, List of Stores Taken North, G.W. Blunt White Library, Mystic Seaport, Mystic, Conn).

aloft at 10 AM bearing SE. We made Nottingham at 2 PM and passed its southern point at 8 PM. Weather improving through the day. Winds NW. Sky overcast, evening clearing.

Friday September 15
Moderate to fresh breezes from NW to WNW with snow squalls. Sky generally overcast. Passed the east end of Charles Island at 4 PM, distance about fifteen miles. Set the squaresail and took in the mainsail. Steering ESE true. Got out flour.

Saturday September 16
Snow squalls with fresh breezes from SSW to WNW. Passed North Bluff at 8 AM. At 2 PM could see the north shore but the weather has been thick most of the time. There are quite a number of icebergs.[2] At 8 PM reduced sail so as to go slow till daylight and also clear the icebergs. Fine snow falling most of the time.

Sunday September 17
Snow squalls and thick weather with strong winds from west. Ran under short sail during the night. There were a number of icebergs all through this day's run. Could not make out either side of the entrance as we came out of the straits at 2 PM. Have passed quite a little field ice till 7 PM, when we came to clear water. Thick fog but now clearing. Wind west true, course east true.

Monday September 18
During the night the wind hauled (backed) [sic] to the SE and has been moderate through the day. Rove off new runners and falls on the main boom lifts. Lat 60°55′N, long 61°03′W. At night calm, sky overcast and misty.

Tuesday September 19
Fresh breezes from NW. Cloudy. Have been making a good run today. At night wind becoming light, the schooner rolling and slatting her sails. Rove off new topgallant braces. The leak is 450 strokes an hour – pump

2 Some of the icebergs calved by glaciers around Baffin Bay are carried westward along the northern shore of Hudson Strait by a current which runs contrary to the general drift of surface waters from Hudson Bay to the Atlantic Ocean. At North Bluff, or thereabouts, the counter-current and its bergs swing around to the left and retreat eastward. Icebergs were a real hazard to whalers entering Hudson Bay in July and leaving it in September.

every hour. One large iceberg passed today. Lat 59°20′N. Long 58°17′W. Distance run last twenty-four hours 135 miles.

Wednesday September 20
The wind backed to the SE and has increased to a fresh gale. A little cold rain. It being directly ahead we are now laying to. Lat 58°25′N, long 56°47′W.

Thursday September 21
A fresh gale during the night from SE. Today more moderate, wind hauling to west-southwest, but there is still an irregular swell quite heavy from south, the vessel pitching quite heavy. Mostly overcast. Lat 57°30′N, long 54°58′W.

Friday September 22
Winds moderating from NW. Lat 55°31′N noon, long 55°05′W. Mostly overcast.

Saturday September 23
Light airs with fog this afternoon. Have passed two large bergs. We are trying to finish drying the bear skins which we got last [August] in Frozen Strait. Today we scraped off what we could of the dry, gummy grease. Gave one to one of the men who has been always ready and willing to help in any way he could. Our cook is sick whenever the sea is rough and this man does his work.

Sunday September 24
Thick fog most of the time with irregular swell and fresh breeze hauling from south to west. Have seen one berg. Lat 54°03′N, long 54°12′W.

Monday September 25
Moderate breezes from south to SE. Partly overcast. Have passed a number of bergs. We are now about forty miles from Belle Isle. At 3 PM lat 52°56′N, long 54°57′W. Got out fresh water and rove off new main throat halyards.

Tuesday September 26
Last night thick cloudy weather. We stood to the SW and at 10 PM raised a flashing white light bearing S by W true, not having such a light down on the charts or in sailing directions. Thought it best to stand off till daylight, which we did, but the weather has been thick. Though we stood

in at daylight [we] gave up trying to go through the Strait of Belle Isle. Have passed several bergs. Very thick fog this afternoon. Wind SSE true. Lit the side lights for the first time. Lat at noon 52°37′N.

Wednesday September 27
Thick fog with light airs hauling from SE to NW during the night. At midnight the air increased from the NW so that we could steer a course southeast by S. A steamer was heard to pass close to us in the fog heading toward the Strait of Belle Isle. Have made 110 miles since midnight till 6 PM. Overcast with some rain. No observations could be got. Running with mainsail tied up. Have seen two bergs this afternoon.

Thursday September 28
The weather has moderated and cleared up, the wind dropping off and backing to the south, then ESE. At 5 PM tacked and we are now steering a course south. Have seen one berg. Could see a light for a few minutes, which we took to be a steamer. It proved to be a star. Could get observations today of the sun and this evening of the North Star. Lat 49°21′N, long 51°21′W. A heavy swell from NW and there seems to be another from the SE.

Friday September 29
Had fresh breezes from SE and east through night and this forenoon. This afternoon light airs from SSW and calms. There is quite a SE swell. Fog and rain during the forenoon, afternoon clearer. Washing the paintwork this morning with potash. Lat by dead reckoning 47°40′N, long by observation 50°57′W. Have sounded twice but no bottom with eighty-five fathoms of line out.

Saturday September 30
Much fog with heavy rain squalls during the night, wind fresh from east to SE. This forenoon the weather improved so that we got more sail on, also became quite clear. The wind hauled to the SW and came out with such force that the jib topsail went to pieces. We shortened sail – fore staysail, foresail and two-reefed mainsail. At 6 PM put one reef in foresail, though the weather is no worse. Could see Cape Ballard, Newfoundland, at 4 PM, distance about fifteen miles. Found that the chronometer is rated all right (by observation). Lat 3 PM 46°45′N, long 3 PM 52°17′W. This seems correct by the land.

Sunday October 1 Off Cape Race, Newfoundland
We have seen the first vessel today since we left – one fishing smack –
and also a large steamer, the steamer bound to the WSW. Quite heavy
weather. Last night we lay to heading to the south. This morning wore
around and could make out the land, probably Cape Ballard. There is a
long swell from the east besides the one from the SW. At noon wore around.
Nine AM lat 46°40′N, long 52°27′W. Heading SW by S, two reefs in mainsail
[with] foresail, staysail and jib. Filled the fresh water cask.

Monday October 2
The wind moderated during the night but the swell keeps up. Today light
airs and calms with a bad swell so that we took in the mainsail but have
done much tossing about. The leak gets worse in such rolling and pitching
– 1,000 strokes per hour. Our cook is unable to work such times so have
to send another man to do his work. At night a little air from NW so that
we have steerage way. Quite pleasant and warm.

Tuesday October 3 Nearly south of Cape Race, Newfoundland
The weather has been quite pleasant with light and moderate winds from
north but backing slowly to west-northwest by night. Saw a three-masted
steamer to the south of us steering SW by W. Have been at work repairing
the jib topsail. Lat 4 PM 45°09′N, long 4 PM 54°09′W. Got out flour.

Wednesday October 4
Moderate breezes from NW to SW. Heading to the SW during the night
and today to the NW. A steamer seen last night bound to the westward.
We passed a broken boat bottom up. It was painted yellow, flat bottomed
and sharp at both ends, about sixteen feet long. Partly clear during the
day but at night overcast. Lat 3 PM 44°40′N, long 3 PM 55°13′W.

Thursday October 5
Moderate breezes from SSW during the night. Today foggy this forenoon
but clear this afternoon with a smoking SW strong breeze. Have one reef
in fore topsail and one in the mainsail. Filled up the water butt. Have
seen five fishing smacks today. Our lat 3 PM 45°05′N, long 3 PM 57°44′W.

Friday October 6
Light and moderate breezes from SW to NW. Today we have seen several
fishing smacks, American (on the bank known as Moraine Bank). Told

Mr Ellis to go on board of one and buy some fish. She proved to be the *Horace B. Parker* of Gloucester, Captain Jesse Martin. Saw one steamer bound east. We got several papers which interest us. The fresh fish are quite a treat to us. Sent letters to the smack to mail. Three PM lat 45°08′N, long 58°49′W.

Saturday October 7
Fresh squalls during the night from NW accompanied by lightning. Today wind fresh but quite steady, at night rather light. Lat 4 PM 44°07′N, long 4 PM 60°59′W. Saw three fishing smacks at anchor on the Middle Ground.

Sunday October 8
We had moderate weather during the night, winds westerly. Today the wind has backed from NW to SW and increased to a fresh gale. Sky quite clear. Lat 4 PM 44°03′N, long 4 PM 61°53′W. Rove off new main topsail halyards. We have two reefs in mainsail and flying jib furled.

Monday October 9
The wind hauled from the SW to NW after midnight. At 3 AM wore around and as the head sea has improved we have made sail. The weather was squally early in the day and there has been a fresh or strong wind all day. Passed an American smack at 6 PM. Lat 4 PM 43°43′N, long 4 PM 63°18′W. Rove off new ——— [fore?] topsail braces.

Tuesday October 10
Had strong breezes from north during the night so that we took in the mainsail and fore topsail. Today more moderate and from the NE. Set the squaresail. This has been the best day's run since we left Repulse Bay. Have seen several fishing smacks. Set up the flying jib stay and topmast and topgallant stays, also topgallant backstays. Our run by log was 185 miles but by observation 205 miles. Lat 42°28′N, long 67°28′W. Sky mostly overcast. Packed up many of my belongings in a cask to take home.

Wednesday October 11
Light breezes and calms during the night. Today the wind has freshened up from the SE, hauled from NE. Have seen a few vessels just at night. Gave the men their slop chest bills. We are in hopes to see the lights at Cape Cod [Mass] this evening. Have not got good sights today, the sky not clearing till this afternoon.

Thursday October 12
Fresh gale from the SE. We made the lights at 11 PM – Nausetts and then Chatham – but the wind blowing so heavy and the weather getting thick we wore around and stood offshore. Rain and fog till 10 AM when the wind hauled to the west. We stood in and made Cape Cod light and have been trying to hold on and work in to the south but there has been quite a swell. We are now close in near Chatham with the wind from the NW. Clear and pleasant. Broke out our last barrel of pork and our sugar is gone and flour nearly so. No beef left. Beans also used up. It is about time we got home.

Friday October 13
We come to anchor near Chatham Light at about 1 AM, the wind being WNW at the time. Today it is west. A large number of vessels have passed bound to the northward. We have had our colors set this forenoon hoping that we might be reported. At 4 PM the schooner *Sagamore* passed close to. Her captain waved his cap and we waved the flag so that I think he will report us if he arrives in Boston tomorrow. He also blew the steam whistle to let us know he understood. Have found that our chronometer instead of being slow one minute and four seconds was fast one minute and twenty seconds. The schooner is leaking 800 strokes an hour laying to an anchor. It is on the increase.

Saturday October 14
Had a fresh breeze from SW during the night and quite a swell with strong tides. This morning weather improving. Got under way with quite a large fleet with the wind moderate from the NW, which backed during the day to SW. We have worked up as far as Vineyard Haven and come to anchor outside the harbor at 9:30 PM. Light winds and head tide.

Sunday October 15
This morning we found we had passed over the Hedge Fence without getting ashore. A Captain Randall came out to us in his little steamer and we made a bargain with him to take us up to New Bedford for half pilotage and twenty dollars for towing. Came to anchor just at sundown. Stayed at Mr Luce's all night and arrived home on the sixteenth, 8:30 PM.

Epilogue

GEORGE COMER

After a winter at home in East Haddam, George Comer set out again for Hudson Bay in the summer of 1906.[1] The schooner *Era* was almost sixty years old but her new owners had caulked the hull and renewed the ice-sheathing. After a few days at sea Comer happily concluded that the ship was 'quite tight,' not making water as she had on the previous voyage; and as she achieved daily runs of 130, 170, and 159 miles, prospects for a fast passage and a successful voyage seemed good. On 27 July, a week or so out from New Bedford, the *Era* was off the south coast of Newfoundland, steering in an easterly direction through fog and rain squalls under close-reefed mainsail. Sun sights had not been obtained for a few days, but soundings appeared to indicate that they were just on the edge of Saint Pierre Bank, a safe distance from land. The *Era* was, however, perilously close to the French island of Petite Miquelon. Just before midnight a lookout suddenly saw breakers ahead, and as Captain Comer tried frantically to wear the vessel around out of danger, she struck bottom hard. Within moments heavy seas were sweeping over the decks. The crew managed to reach shore but the *Era* was wrecked, ending a distinguished career during which she had completed twenty whaling voyages to the Arctic, coping successfully with all the hazards of that intemperate region only to run ashore in a Newfie fog.

George Comer was not so easily destroyed, however. In 1907 he was

---

1 The brief synopsis of this voyage is based on George Comer, Manuscript Journal on Board the *Era* 1906, G.W. Blunt White Library, Mystic Seaport, Mystic, Conn.

given command of the 86-ton schooner *A.T. Gifford* of Stamford, Conn, took her north on a two-year voyage into Hudson Bay, and on another long cruise in 1910–12. In 1915 he accepted the position of ice pilot on the *George B. Cluett*, a schooner chartered by the American Museum of Natural History to sail to northwest Greenland and pick up MacMillan's exploration party, which had been two years in the north. The relief vessel was trapped in the ice beyond Cape York and Comer spent the next two years in Greenland, assisting the museum staff by important archaeological investigations in what came to be known as 'Comer's midden.'[2] Following his return in 1917, he enlisted in the United States Naval Reserve and made voyages to Europe and South America. He returned to Hudson Bay for the last time in 1919, when he sailed as master of the yacht *Finback* on a private expedition under Christian Leden to combine ethnological research with commercial trading among the Eskimos of western Hudson Bay. Unfortunately (and ironically) this venture, the last voyage of Comer's career, ended in shipwreck at the place so familiar to him, Cape Fullerton.

The last eighteen years of his life were spent in retirement, but not in inactivity. In 1920 he was the first witness to appear before the Royal Commission in Ottawa investigating the feasibility of establishing herds of reindeer and musk-oxen in the Canadian Arctic and subarctic. A few years later he took the position of master on the schooner *Blossom*, to sail on a two-year scientific expedition to the south Atlantic and Indian oceans, but resigned at the last minute, perhaps fortunately, for the voyage turned out to be unhappy, marred by poor leadership and crew discontent. Lectures before a number of clubs and organizations kept him busy, and he exchanged ideas with scientists at every opportunity. Active in community affairs, he served a term in the Connecticut legislature. In his seventy-ninth year George Comer was admitted to the Veterans' Hospital

2 Comer's midden at Umanaq quickly became one of the most significant archaeological sites in arctic North America and Greenland. Its rich assemblage of artifacts, with an abundance of baleen and whale bones, was considered typical of what was later called the Thule Eskimo culture, and for a time it was believed to represent the earliest inhabitants of Greenland. The material excavated by Comer in 1916 went to the American Museum of Natural History, where it was analysed by Clark Wissler, and the artifacts unearthed by Danish archaeologists were deposited in the National Museum at Copenhagen. During the early 1920s the work of the Fifth Thule Expedition in the Canadian Arctic revealed other sites remarkably similar, notably that of Naujan at Repulse Bay. The belief in the antiquity of the Thule culture was shattered about 1925, however, by Diamond Jenness's discovery of the older Dorset culture along the north shore of Hudson Strait (Frederica de Laguna, personal communication).

in Northampton, Mass, suffering from Bright's disease. Two weeks later, on Tuesday, 27 April 1937, he died.

THE ESKIMO

Albert Peter Low, the geologist in charge of the *Neptune* expedition of 1903-4, remarked that 'a withdrawal of the whalers would lead to great hardship and many deaths among these people if the Government did not in some manner take their place and supply the Eskimos with the necessary guns and ammunition.'[3] Without a regular supply of ammunition their guns would be useless and they would have to return to the primitive weapons formerly used. But was that possible? Liberally supplied with guns from the whalers (a census in 1911 showed that each family in Arctic Bay, Baffin Island, owned an average of 2.6 guns),[4] they had abandoned the manufacture and use of bows, arrows, lances, and spears. It was too much to expect that a people who had enjoyed for several decades the sophisticated weapons of an external civilization could revert to the old ways.

Fortunately, however, whaling was succeeded by organized trading activities, at first by small Scottish, English, and Canadian firms, but increasingly by the large and powerful Hudson's Bay Company which, after two-and-a-half centuries of fur trading south of the tree-line, was at last eager to take advantage of the resources of the arctic regions. As year-round trading posts were established in the wake of the whalers the Eskimos were able to continue to obtain the material goods they required in exchange for furs, so that the crisis predicted by Low happily did not come about. In time the inauguration of annual government ship voyages to the arctic settlements, the erection of missions and more police posts, and the extension into the north of a number of administrative agencies and commercial activities gradually brought to the native population the benefits of health services, education, modern communications, social welfare, employment opportunities, municipal government, and co-operative organizations.

These developments could not have occurred if the dispersed Eskimo population had not become concentrated in reasonably large settlements accessible by ships and aircraft. The last three decades have been char-

3 A.P. Low, *The Cruise of the Neptune, 1903–4*, p 271.
4 'Census of the Far North of Canada,' *Fifth Census of Canada 1911* (Ottawa: King's Printer 1913), appendix, 643.

acterized by demographic centralization as well as rapid cultural change. The centralization had its roots in the whaling period. Eskimos were attracted to wintering ships because they could obtain jobs, material goods, food, and some medical attention, and because they could enjoy the stimulus and excitement of a large gathering with a diversity of people and activities – much the same combination of practical and intangible factors that have been drawing rural dwellers into urban communities throughout the world. And the whaling captains, like today's administrators, could not provide such services unless the Eskimos congregated near the winter harbours.

In northwestern Hudson Bay there are now more Eskimos than in Comer's day, owing to natural increase and the relocation to coastal regions of inland groups whose resource base had diminished. The old tribal or territorial concepts have faded and people of various origins mix together in the towns of Repulse Bay, Coral Harbour, Chesterfield Inlet, Baker Lake, and Rankin Inlet. They live in oil-heated houses laid out in streets, and have running water, toilets, and modern kitchens. Sitting on comfortable chairs and sofas they passively watch 'MASH' and 'All in the Family' on colour television beamed in by satellite. The men drive snowmobiles, operate freight canoes with large outboards, and a few own cars to drive on the mile or two of roads around town. Teenagers listen to rock music, wear jeans and fake cowboy boots, and delight in raising dust and noise with Japanese trail bikes. Children play briefly with the usual assortment of ugly plastic toys and discard the broken bits and pieces everywhere. The women use sewing machines and clothes washers, receive regular family allowance cheques, and pick up their mail-order World of Beauty kits at the post office. Some people work at eight-to-five jobs and others are on unemployment insurance. Everyone shops at the large general store of the Hudson's Bay Company for food, clothes, and most other commodities.

In the absence of a written language the traditional Eskimo passed on information orally from one generation to the next. The American explorer C.F. Hall was astonished to hear an account of Frobisher's visit almost three hundred years previously, and George Comer recorded Iglulingmiut songs describing Parry's ships in Foxe Basin in 1822.[5] But the adoption

5 Iglulik songs about the arrival of Parry's two ships in 1821, and the thoughts of the Eskimos on seeing them, were recorded by Comer about eighty years after the event. They are preserved on wax cylinders 1108 and 1112 at the Folklore Institute, Indiana University, Bloomington, Ill. Unfortunately, they are almost totally incomprehensible owing to the deterioration of the cylinders.

of a southern culture, with its extraordinary emphasis on visual communication has dealt a crippling blow to oral transmission of knowledge and tradition. There, as here, the younger generations are largely oblivious of the heritage of the past and not much interested in hearing or reading about it. Vivid images of the whaling days live yet in the memories of a handful of old men and women, but they may not live for long.

WHALING

The gestation period of a bowhead whale is approximately ten months. A cow bears only one calf at a time, which must be suckled for almost a year, and she normally gives birth not more than once every two years. Four years may pass before the calf becomes sexually mature, if indeed it survives that long, for almost half the calf crop may be lost in the first year because of birth complications, disease, parasites, wound infections, killer whale predation, groundings in shallows, or the rapid freezing over of the sea. As a result, the whale population grows slowly; it cannot withstand severe hunting pressure.

Whatever the distribution of Hudson Bay whales in winter (even today not known accurately) it is clear that from May to September they tended to congregate on summer feeding grounds in northwestern Hudson Bay. There they were partly confined within the narrow space of Roes Welcome Sound and the restricted waterways of Frozen Strait, Hurd Channel, and Lyon Inlet. Their movements were limited even further by the presence of pack and landfast ice early in the season. In this geographical setting the whales were highly vulnerable to human predation. For half a century American and British whalemen, with their Eskimo helpers, killed as many whales as they were able. They took more than 300 in the first decade, averaging approximately six whales per voyage, but such intense pressure could not be sustained by the whale population.[6] Average yields declined to four whales per voyage in the second decade and dropped below two in the third. As yields diminished so did the number of vessels in the fishery. Whaling was finished by 1915.

6 The catch data here are taken from W. Gillies Ross, 'The Annual Catch of Greenland (Bowhead) Whales in Waters North of Canada 1719–1915: A Preliminary Compilation,' 144, Table 4. The numbers of voyages per decade are from W. Gillies Ross, *Whaling and Eskimos: Hudson Bay 1860–1915*, 37. It should be borne in mind that most Hudson Bay voyages were wintering ones which incorporated at least two whaling seasons. The average catch per ship-season (rather than per voyage) works out to 4.4 in the first decade 1860–70, 2.7 in the second, and 1.1 in the third.

As whale returns dwindled the whalemen diversified their operations, obtaining furs, skins, and ivory in trade from the natives. By the time of Comer's last whaling voyage fur trading was of paramount importance; the *A.T. Gifford* returned in 1912 with the skins of 412 fox, 133 polar bear, 114 musk-oxen, 81 wolf, and 34 wolverine.[7] The *Gifford* was fortunate in capturing five whales as well, but their baleen measured 3'6", 4'2", 4'6", 6'5", and 8'2", a pathetic contrast to the large, mature whales of earlier years, whose bone often exceeded ten feet in length.

For three centuries there had been little concern for the survival of the species as stocks of Greenland, or bowhead, whales were systematically reduced throughout the circumpolar north. But when the arctic whaling industry had irresponsibly destroyed the renewable resource upon which it was based – when it was no longer profitable to send out whaling ships – then the whalemen, the companies, and the governments self-righteously espoused the cause of conservation. International legislation in 1931 imposed a ban on the killing of *Balaena mysticetus*, but was it to save the species or to save face? The recovery of bowhead stocks was confidently predicted, but in the intervening half-century only a few dozen sightings of the whales have been reported in Hudson Bay, and there are some who think that it is no longer within the power of the stock to recover, or even to survive.

Sadly, the lessons of the past are often ignored. When the era of modern whaling began after the First World War, with a level of technology making possible the capture of all the large cetacean species, including the majestic blue whale, the economic philosophy (if it can be dignified as such) of the whaling industry was unchanged: kill as many whales as catchers and factory ships could manage and industry could use, paying only token attention to the principle of sustainable yield. Now, less than three-quarters of a century after the collapse of the bowhead whale fisheries in the regions of Davis Strait, Hudson Bay, and Bering Sea, several other species of great whales have been brought to the edge of extinction. Such is the legacy of whaling.

7 George Comer, Manuscript Journal on Board the *A.T. Gifford* 1910–12, G.W. Blunt White Library, Mystic Seaport, Mystic, Conn, end papers.

The schooner *Era*

During the nineteenth century larger and more powerful vessels were developed for the arctic whale fisheries. Larger ships, which carried more supplies and more oil and bone, could remain out longer and waste less time on voyages back to port. More powerful whalers, with steam auxiliary engines and propellers to drive them through ice, could cruise more extensively in arctic waters and increase the duration of the whaling season by breaking out of winter harbour earlier in spring and returning later in fall. In addition, engines could extricate a vessel from perilous situations when wind and currents were unfavourable. The Scots pioneered the use of steam in the Davis Strait fleet in 1857 and soon proved the merits of auxiliary power. When American whalemen penetrated into the western Canadian Arctic in 1889 they made effective use of steam whalers, the largest of which – the *Orca* – was 628 tons.[1]

Curiously, in Hudson Bay vessels became smaller rather than larger, going from an average size of over 200 tons in the first decade of whaling (1860–70) to less than 170 in all subsequent decades and as low as 125 in 1880–90. Only two whalers, one American and one Scottish, were equipped with engines. All the others relied on wind alone. They suffered the consequences; one out of every ten voyages ended in shipwreck.

Among the undersized vessels of Hudson Bay, schooners comprised more than a third of the fleet during the half-century of whaling in the region. Barks (the most common rig), brigs, and ships made up the rest. Schooners were the smallest, averaging slightly over 100 tons, compared to about 160 tons for brigs, 225 tons for barks, and 400 tons for ships.

1 John R. Bockstoce, *Steam Whaling in the Western Arctic* (New Bedford, Mass: Old Dartmouth Historical Society, 1977), 86.

Whereas large steam whalers could use their power to butt through ice and their speed to reach the whales and avoid danger, the small sailing schooners had to depend on their shallow draft, their ease of handling, and their manoeuvrability to survive near the hazardous arctic coasts.

Built in Boston in 1847 the schooner *Era* was employed in the coastal trade until 1864, when she was sold to the firm of Moses Darrow in New London and made her first arctic whaling voyage, to Baffin Island under Captain Bellows. After another trip under Bellows, in which the vessel wintered in Cumberland Sound, she was sold to Williams and Barnes, also of New London, and completed a number of whaling voyages to Cumberland Sound, Hudson Strait, and Hudson Bay under captains Tyson, Miner, Spicer, and Clisby. In 1895 the *Era* was bought for $1,600 by Thomas Luce of New Bedford, who installed Comer as master and sent the schooner on a series of voyages to Hudson Bay. By 1903 the *Era* had made nineteen voyages to the Canadian Arctic and had spent twelve winters frozen into the ice. On the way to Hudson Bay in 1906 she went ashore and was lost.[2]

The *Era* was a two-masted topsail schooner, originally 188 tons but after alterations some time between 1870 and 1877, only 134 tons. As the figure shows, both foresail and mainsail were gaff sails with booms. All the headsails were loose-footed. Topsails were used on both foremast and mainmast; those on the foremast were squaresails. Whaleboats hung from three pairs of davits, two on the port side and one on the starboard side, aft. The tryworks were on the upper deck just abaft the foremast. Blocks were attached to the mainmast for cutting in whales along the starboard side amidships. In winter the main boom was used as a ridgepole for the house built over the after part of the vessel.

2 Biographical details on the *Era* are taken mainly from Barnard L. Colby, 'New London Whaling Captains,' *Publications of the Marine Historical Association* (Mystic Seaport, Conn 1936), vol 1, no 11; Alexander Starbuck, *History of the American Whale Fishery from its Earliest Inception to the Year 1876* (1878; reprint, New York: Argosy - Antiquarian 1964), vol 2; Reginald B. Hegarty, *Returns of Whaling Vessels Sailing from American Ports: A Continuation of Alexander Starbuck's 'History of the American Whale Fishery' 1876–1928* (New Bedford, Mass: Old Dartmouth Historical Society 1959).

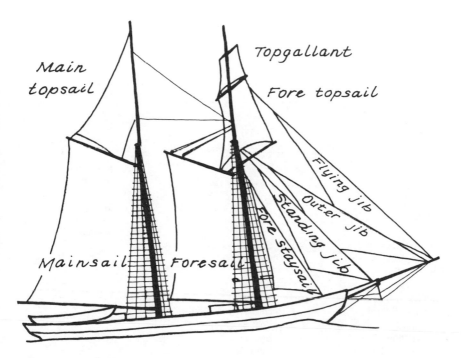

Main topsail

Topgallant

Fore topsail

Flying jib

Outer jib

Standing jib

Fore staysail

Mainsail Foresail

The schooner *Era*

APPENDIX B

Crew list of the *Era* 1903

Name	Origin*	Rank
Comer, George		master
Ellis, Richard L.		first mate
Suares, Jules†		second mate and boatsteerer
Reynolds, Herbert R.‡	Brockton	third mate
Tucker, Charles		boatsteerer
Soares, Manuel Jose		"
Lopes, Brass		"
Samuel, Alexander		steward
Evans, George F.	Lynn	seaman
Greenlau, Harry B.	Calais, Maine	"
Bassett, Charles B.	Old Orchard, Maine	"
Robbins, Byron B.	Boston	"
Khitarion, S.M.	Boston	"
Webster, Charles	Readville	"
Neale, William J.	New Haven, Conn	"
Strassburg, Jacob L.	Boston	"
Cleary, James J.	Brockton	"
Huse, Malcolm	Georgetown	"
Cohl, Augustus A.	Boston	"
Maynes, W.P.	Boston	"

SOURCE: undated newspaper clipping pasted into George Comer, Manuscript Journal on Board the *Era* 1903–5 (G.W. Blunt White Library, Mystic Seaport, Mystic, Conn), vol 3, title page.

* It appears that, unless otherwise stated, Massachusetts was the place of origin.
† Evidently failed to appear at time of departure.
‡ Apparently promoted to second mate to replace Suares.

Stores carried on the *Era* 1903

PROVISIONS

84$^5/_8$	barrels flour	5	pounds mustard
21,803	pounds bread	2	pounds cinnamon
27$^1/_2$	barrels beef	2	pounds summer savory
22	barrels pork	15	pounds pepper
300	pounds rice	10	pounds ginger
5	barrels meal	2	pounds sage
1,900	pounds coffee	5	pounds hops
130	pounds tea	161	pounds saleratus
1	barrel vinegar	3	dozen baking powder
1,202$^1/_2$	pounds butter	12	pounds cream of tartar
5	boxes soap	5	dozen Magic Yeast
504	pounds dried apples	50	bushels beans
1,982	pounds sugar	5$^1/_2$	bushels peas
504	pounds canned corned beef	12	pounds chocolate
		100	pounds raisins
35$^1/_2$	dozen clams	6	boxes salt
24	dozen tomatoes	150	pounds peanuts
12$^1/_4$	dozen milk	100	pounds codfish
32	dozen corn	1,040	pounds evaporated potatoes
12	dozen pears		
3	barrels potatoes	1	dozen Hires root beer
3	bags onions	1,202$^1/_2$	gallons molasses

SOURCE: George Comer, Manuscript Journal on Board the *Era* 1903–5 (G.W. Blunt White Library, Mystic Seaport, Mystic, Conn), vol 4, 208–13.

100	pounds dried peaches	75	pounds evaporated
1	pound nutmegs		peaches
2	pounds cloves	6	casks fresh water (1,496
3	casks cabbage (pickled)		gallons) and
			scuttlebutt full

EQUIPMENT AND MATERIALS

1	coil deep sea line (new)	12	axes
1	coil $2^3/_4$ manila	100	pounds $^1/_4$ and $^3/_8$ rod
1	coil $2^1/_2$ manila		iron
1	coil $2^1/_4$ manila	50	pounds $^1/_2$ rod iron
1	piece 8-ounce	5	pounds blue
	_____[raven?]	20	pounds whiting
1	bolt no 1 duck	10	pounds putty
12	pieces_____[boat?] sail	$1^1/_2$	kegs black (50 pounds)
2	pounds hemp twine	10	pounds black in can
93	sail needles	450	pounds white lead
5	pounds wax	1	barrel linseed oil, raw
10	pounds sail twine	2	gallons turpentine
1	can tacks	2	gallons dryer
15	gallons coal tar	5	packages sandpaper
22	window frames	2	packages emery cloth
50	panes glass $9^1/_2 \times 13^1/_2$	9	pounds copper tacks
2	8-inch hack saws	$40^1/_2$	tons coal
48	8-inch blades	430	gallons kerosene
3	gross assorted screws	2	double barrel shotguns
2	pounds assorted	2	single barrel shotguns
	washers 1 to $^1/_4$	2	sets loading tools
100	pounds ten-penny wire	100	paper shells
	nails	100	brass shells
50	pounds eight-penny wire	35	gross matches
	nails	2	boxes pipes
20	pounds thirty-penny wire	2,040	pounds tobacco
	nails	12	packs playing cards
20	pounds sixty-penny wire	2	rugby footballs
	nails	2	baseballs
10	pounds $3^1/_2$-inch wrought	1	3-inch skillet
	nails	1	5-inch skillet

20	pounds 3-inch wrought nails	6	oil stoves for boat
10	pounds 2½-inch wrought nails	6	tea kettles
5	pounds 2-inch wrought nails	6	frying pans
		6	can openers
		24	whetstones
		10	square point shovels
30	pounds boat nails	72	iron staples

LUMBER

600	feet cedar boards
100	feet 3 × 4 scantling
275	feet 2 × 4 scantling
330	feet 2 × 8 sled plank
65	feet ash
2,303	feet white pine for house over vessel
92	feet hard pine

WHALING GEAR

72	darting lances
75	shooting lances
113	shells
191	wads
9	darting guns
4	Egger shoulder guns
2	shoulder guns
1	Brand shoulder gun
25	lances
16	coils tow line
35	toggle irons
52	gun irons
48	rowlocks
12	15-foot oars
12	16-foot oars
4	17-foot oars
2	18-foot oars
6	steering oars
14	boat booms
12	boat masts

26	spare poles
1	cask flagging
1	cask sawdust
1	turning lathe

S L O P S

20	suits oil skins
22	sou'westers
72	undershirts
72	pair drawers
72	overshirts
30	duck coats
31	pair blankets
36	tin pots
36	tin pans
36	iron spoons
24	caps
24	pair brogans
7	pair slippers
95	pair mittens
108	pair stockings
4	razors
84	pair pants
12	pieces mosquito netting
12	boxes thread
	woolen yarn
20	_____[sheath?] & belts
4	pieces coverlet cloth 181$\frac{1}{4}$ yards
1	piece denim 52 yards
19$\frac{1}{2}$	gross buttons (miscellaneous sizes)
6	dozen cotton thread
6	dozen combs
48	towels
72	handkerchiefs
2	gross thimbles

TRADE GOODS

Articles for hunting and fishing
16 cans powder
24 38/40 no 2 Remington rifles
12 sets loading tools
12,000 primed shells 38/40
1,000 primed shells 38/40 straight
200 45/70 primed shells
75,000 no 1 Winchester primers
10,000 no 2 Winchester primers
5,000 Army caps
5,000 percussion caps
2 quarts of flints
600 Limerick fish hooks, no 8
600 Kerby fish hooks, no 8
10 pounds fish line

Tools and general implements
100 jack-knives
60 sheath knives
36 12-inch snow knives
18 boat hatchets
18 shingling hatchets
26 cross-cut saws
10 10-inch steels
12 8-inch mill bastard files
12 10-inch mill bastard files
6 6-inch three-cornered files
48 assorted gimlets
24 handled awls
9 ripping saws

Domestic articles
24 match boxes
12 sets dominoes
25 pounds beads
731$^{1}/_{2}$ yards calico
6,000 needles

5,000 glover's needles
24 large forks
24 large spoons
74 pairs scissors
48 chopping trays
48 chopping knives

Beaufort wind scale

Force on Beaufort scale	Nautical miles per hr	Description	Height of sea in ft	Deep sea criteria
0	0–1	calm	—	flat calm, mirror smooth
1	1–3	light airs	$^1/_4$	small wavelets, no crests
2	4–6	light breeze	$^1/_2$	small wavelets, crests glassy but do not break
3	7–10	light breeze	2	large wavelets, crests begin to break
4	11–16	moderate breeze	$3^1/_2$	small waves, becoming longer, crests break frequently
5	17–21	fresh breeze	6	moderate waves, longer, breaking crests
6	22–7	strong breeze	$9^1/_2$	large waves forming, crests break more frequently
7	28–33	strong wind	$13^1/_2$	large waves, streaky foam
8	34–40	near gale	18	high waves of increasing length, crests form spindrift
9	41–7	strong gale	23	high waves, dense streaks of foam, crests roll over
10	48–55	storm	29	very high waves, long overhanging crests; surface white with foam

Force on Beaufort scale	Nautical miles per hr	Description	Height of sea in ft	Deep sea criteria
11	56–65	violent storm	37	exceptionally high waves, sea completely covered with foam
12	above 65	hurricane	—	air filled with spray, visibility seriously affected

SOURCE: Peter Kemp, ed, *The Oxford Companion to Ships and the Sea* (London: Oxford University Press 1976), 72.

Minimum air temperatures at Fullerton Harbour (daily minimum temperature − 40°F or below) 1903–4

16 January (1904)	− 40°F
17 "	− 41
18 "	− 41
20 "	− 42
30 "	− 40
31 "	− 40
1 February (1904)	− 42
2 "	− 43
3 "	− 42
19 "	− 46
20 "	− 44
24 "	− 42
25 "	− 46
26 "	− 46
27 "	− 43
28 "	− 46
29 "	− 45
1 March (1904)	− 46
2 "	− 52
3 "	− 53
4 "	− 40
5 "	− 41
25 "	− 40
26 "	− 41
27 "	− 41

SOURCE: A.P. Low, *The Cruise of the Neptune 1903–4*, appendix I, 300–13.

Sea ice thickness at Fullerton Harbour

Winter 1903–4		Winter 1904–5	
Date	Thickness (inches)	Date	Thickness (inches)
		Oct 24	9
		31	$14^1/_2$
Nov 16	13	Nov 7	$14^1/_2$
23	—	14	15
30	21	21	19
		28	23
Dec 7	24	Dec 5	23
14	—	12	$23^1/_2$
21	—	19	31
28	31	26	32
Jan 4	33	Jan 2	$33^1/_2$
11	38	9	$35^1/_2$
18	$39^1/_2$	16	39
26	41	23	43
		30	44
Feb 1	45	Feb 6	51
8	48	13	$48^1/_2$
15	50	20	52
22	53	27	57
29	55		
March 7	58	March 6	58
14	62	13	56
21	64	20	63
28	68	27	72
April 4	70	April 3	73
11	71	10	74
18	73	17	74
25	74	24	75

239 Sea ice thickness at Fullerton Harbour

Winter 1903–4		Winter 1904–5	
Date	Thickness (inches)	Date	Thickness (inches)
May 2	74	May 1	75
9	$70\frac{1}{2}$	—	—
16	$71\frac{1}{2}$	—	—
23	75	20	$64\frac{1}{2}$
30	72	—	—
June 6	70	June 3	61
13	66	10	60
19	58	17	47
27	45		
July 4	22 to 45		
11	30		

SOURCE: Figures are from George Comer, Manuscript Journal on board the *Era* 1903–5 (G.W. Blunt White Library, Mystic Seaport, Mystic, Conn), vol 3, 77, and vol 4, 25, except for the data from 23 May to 27 June 1904, which are from A.P. Low, *The Cruise of the Neptune 1903–4*, appendix I, 309–11. All the measurements appear to have been taken by the scientific staff of the *Neptune* and *Arctic* and copied into Comer's journal at the end of each calendar year.

Details of whales killed in 1905

Date	Place	Boat	Whale Sex	Bone Length	Bone Weight (pounds)	Number of bundles
June 13	C Kendall	Mr Reynolds	Bull	9'6"	1,456	18
June 16	Whale Pt	Sam	Cow	4'6"	241	
July 30	Lyon Inlet	Gilbert	Bull	6'10"	612	12
Aug 1	Lyon Inlet	Mr Reynolds	Cow	6'10"	524	9
Aug 4	Lyon Inlet	Starboard*	Bull	9'0"	1,397	15
Aug 4	Lyon Inlet	Mr Ellis	Cow	8'10"	1,136	24
Aug 13	Lyon Inlet	Mr Reynolds	Bull	9'3"	1,381	19
Aug 17	Lyon Inlet	Starboard	Cow	6'3"	462	11
Total†					7,209	108
Sent to Scotland last year‡					1,853	
Total					9,602	

SOURCE: George Comer, Manuscript Journal on Board the *Era* 1903–5, (G.W. Blunt White Library, Mystic Seaport, Mystic, Conn), vol 4, 214.
* The starboard boat was headed by Captain Comer himself.
† The bone of the eight whales averaged 7'7" in length and 901 lbs in weight. After cleaning and drying it was tied into 108 bundles, averaging 67 lbs. each.
‡ The bone sent out on board the *Active* on 8 September 1904 was from the two whales secured by the *Era*'s native crews at Repulse Bay (see journal entry 8 August 1903) and the small whale killed by native Harry on 15 July 1904. The total yield was probably greater than 1,853 lbs; Comer believed that 100 slabs had been removed from the cache at Repulse Bay (see entry 20 August 1903).

Whaling and trading returns 1905

According to Hegarty's statistical summary of American whaling voyages, the *Era* returned in 1905 with 25 barrels of oil and 6,500 pounds of bone.[1] Comer's journal, however, records 1,853 pounds of bone sent to Scotland on the Dundee whaler *Active* from three whales killed in 1903 and 1904, plus another 7,209 pounds obtained from eight kills in 1905, a total of 9,062 pounds (see appendix G). Hegarty's summary probably overlooks the amount sent by way of Scotland, and the discrepancy between his figure of 6,500 and Comer's figure of 7,209 may be in part a result of weight shrinkage after initial weighing of the bone on board ship, through subsequent processes of cleaning, scraping, and drying.

Fur returns are not included in Hegarty but journal entries for 17 May 1904 and 28 April 1905 list furs taken to date, and the end pages of the journal (vol 4, 215) contain a 1905 inventory of furs packed away in casks, which probably represents the total fur catch of the voyage. The amounts are given in Table H1.

The *Era* also carried furs from the company's station at Wager Bay, namely 350 musk-ox, 16 polar bear, 14 wolf, and 6 wolverine (see journal entry 14 September 1904).

The value of the cargo, including whale products and animal furs, is said to have been $75,000.[2] Whalebone, which brought an average price

---

1 Reginald B. Hegarty, *Returns of Whaling Vessels Sailing from American Ports. A Continuation of Alexander Starbuck's 'History of the American Whale Fishery' 1876–1928* (New Bedford, Mass: Old Dartmouth Historical Society 1959), 36.
2 Barnard L. Colby, 'New London Whaling Captains,' *Publications of the Marine Historical Association* (Mystic Seaport, Mystic, Conn 1936), vol 1, no 11, 223.

Table H1. INVENTORY OF FURS

	May 1904	April 1905	End of voyage
Musk-ox	101	130	134
Fox	96	164	172
Wolf	26	61	61
Polar bear	15	40	50
Wolverine	4	11	14

of $5.80 per pound in 1904 and $4.90 per pound in 1905, was the most valuable part of the cargo.[3]

In addition to the animals secured for the commercial fur trade many were killed for food, oil, and skins for clothing. It is impossible to reconstruct the numbers accurately but the numerous references in Comer's journal provide minimum figures. He records the kills of, or the receipt of meat and skins from, approximately 220 caribou, 180 seal, 30 bearded seal, 25 walrus, 200 ducks, 230 salmon, 80 hares, 25 ptarmigan, and a few geese, as well as more than ten dozen eggs of gulls, ducks, geese, and swans.

3 Charles H. Stevenson, 'Whalebone: Its Production and Utilization,' *Bulletin from Johnny Cake Hill*, Old Dartmouth Historical Society (Winter 1965–6): 8 (originally published by the Department of Commerce and Labor, Washington, Bureau of Fisheries Document no 626, 1907).

Population of Eskimo groups

George Comer compiled population data on the various Eskimo groups by making head counts among groups residing at the winter harbours and by interviewing visiting members of groups inhabiting remote regions. His figures provide the earliest detailed information on Eskimo numbers and sex ratios in the central Arctic. The importance of his data was clear to the anthropologist Franz Boas, who reproduced them in his publications of 1901 and 1907, and to A.P. Low, who included some of the data in his 1906 report on the *Neptune* expedition. There are slight discrepancies among the population figures given in his various journals and published papers, but they are of only academic concern. Table I1, based on Comer's data, lists the men, women, and children in each of the groups represented at Fullerton Harbour in 1903–5, according to what appear to be the least ambiguous and most reliable figures.

Table I1.   Approximate levels of population by sex and group

Group	Men	Women	Boys	Girls	Total
Aivilingmiut 1898*	26	34	27	15	102
Aivilingmiut 1908†	29	42	29	25	125
Qaernermiut 1898* (Kenepetu)	35	46	38	27	146
Qaernermiut 1908‡	38	55	30	42	165
Netsilingmiut 1902§	119	123	138	66	446
Sauniktumiut 1902§ (Show-vock-tow-miut)	46	58	41	33	178

* Franz Boas, 'The Eskimo of Baffin land and Hudson Bay,' 7.
† George Comer, 'Number of Iwilic natives February 10, 1908.' Manuscript list, George Comer Papers, East Haddam, Conn.
‡ George Comer, 'Number of men, women and children in the Kenepetu or Kiackennuck tribe' (1908). Manuscript list, George Comer Papers, East Haddam, Conn.
§ Franz Boas, 'Second report on the Eskimo of Baffin Land and Hudson Bay,' 377–8.

General observations by George Comer[1]

FREEZING A VESSEL IN

With regard to freezing in and the effect it has on a vessel where the ice freezes to something like six feet thick and then great drifts will form (especially when there are native igloos near by), I think it strains them and the weaker a vessel is the more she is affected by the pressure from the ice, as it does not settle evenly but naturally more where the drifts are greatest. It is a common thing to hear quite loud reports and feel the vessel jar during the winter. After a vessel becomes knit to the ice and then large drifts form around, the ice will settle and carry the vessel down with it, then the water will leak through to the top, which freezes and makes harder ice from mixing with the snow.

The vessel in a small way helps to hold the ice up. In the spring of 1896 (February 19) the vessel came up out of the ice something over a foot – the drifts had become so heavy. No harm came to the rudder but certainly the strain must have been great in the pintals. February 19, 1898, we cut both ends clear but the vessel came up slowly at first, then afterwards took a jump. This year 1899 we cut the ends clear the second of March. The vessel has come up very slowly aft but shows no sign of it amidships. She has come up three inches aft. It may be the great weight of snow has settled the ice, or it may be partly that the vessel has been strained and is now coming back straight.

With regards to the ice breaking up in spring ... I have thought by

---

1 George Comer, Manuscript Journal on Board the *Era* 1897–9 (G.W. Blunt White Library, Mystic Seaport, Mystic, Conn). The editor provided the title but the subheads are Comer's own.

experience that instead of putting all the ashes in a heap, as is customary, that a better way is to spread them out thinly in lines leading from the vessel to the shore ahead, and leading from each bow. In the spring they will work down through the ice and when the ice starts out, as it sometimes does, in large bodies, the ice being cut through in several places allows the ice to go by. We tried this last winter though it did not help us that season for the ice broke up with a gale blowing in from the east. If a vessel was going to be towed out early this would not help or be any use.

We are spreading the ashes out in lines again this winter. One of the great faults of this place is [that it is] so large, though quite landlocked. The bottom is quite uneven and generally rocky, varying from six fathoms to sixteen. In one part of the bay it is thirty fathoms deep and from where we are to the head of the bay is from twelve to fifteen miles, making a bend and gradually narrowing up till the head is reached, into which a stream flows in the summer time, where the natives catch many salmon (or salmon trout).

GUNS AND AMMUNITION

The rifles should be breech-loading and single shot such as the Springfield Rifle. A Winchester is more likely to get out of order and become useless. The calibre should not be larger than .44 and perhaps not less than .38. Larger sizes take so much more lead in reloading, also more powder, which is quite an item with the natives after they own the guns. Cartridges should be loaded when about to be needed. Should advise carrying empty shells; if taken from home loaded they are likely to become corroded and will not stand reloading but a little. Shotguns I would recommend a number 10 bore (single barrel), also brass shells to be reloaded. There are large quantities of eider ducks to be seen and large bored guns and BBB shot should be used in order to make hunting them successful.

Large knives if could be had which were *not* tempered high would be better in this extreme cold. They would not be so likely to break.

Revolvers are of no use whatever. Perhaps a heavy revolver would be used sometimes in hunting musk-ox.

TRADE FOR THE NATIVES

There is much that the natives like but there are comparatively few things that are useful to them, such as rifles, powder, lead, _____ . All guns should be breech-loading center fire. Knives with blades ten to twelve inches long are very useful in cutting out blocks of snow in building their igloos (snow houses). In summer-time they use skin tents, a few bought tents. Striped goods would be good trade with the natives. Large and small chopping knives or what is known as saddlers' knives are very useful with the women. Fish hooks and fish lines are also good. Files and saws, also hatchets. Small telescopes are very useful to the men when hunting and are prized very much. Needles should not be forgotten, large and small (but neither extreme). Pants, buttons, also calico and buttons to match, though not actually necessary. Combs and scissors are good. The squaws have a great passion for small glass beads of different colors. A very useful thing is a long square tin about five inches deep with a ring at each corner to hang it up by over the lamp to melt snow in and also cook in. It should have a cover. Matches and tobacco should not be forgotten. Another thing that should not be forgotten is planks for sled runners. These should be twelve or fourteen feet long and eight inches deep then with a shoeing of bone or oak or yellow pine one inch thick and three-and-a-half inches wide.

WATER

With regards to the water in the ship, there is no danger of its freezing in the casks until well along in the winter. Have known where a cask was overlooked and did not freeze until March.

In cutting ice on a pond would recommend selecting a pond of large size or where one has a good depth of water. We have cut ours these winters (1895–99) on a pond that was quite shallow. The water does not seem to be so clear as it ought to be, being of a milky color. The ice is also a dull white color. The water is probably alright as up to now (January 15) no one has been sick on either vessel (there are two here, bark *Canton* and our schooner *Era*. Lat 64°00′ N, Long 88°44′ W – Cape Fullerton, Hudson Bay). The ice was cut about the twelfth of October when it had become about eight inches thick, cut in squares two-and-a-half by three feet and set up on edge close to the shore of the pond. Cut 800 cakes, which is to last through the winter, and have enough to melt down to last us home. Was much more than enough.

One case of scurvy broken out in the *Canton* on one of the men, a Portuguese sailor, about January 15. Later this man died about the first of August and was buried at sea. January 1899: we are now wintering here for the third winter and so far have had no sign of scurvy. A few cases of scurvy on board the *A.R. Tucker* last winter but no one died.

FISH AND GAME

There are large numbers of eider ducks here during the winter and good shooting can be had if one can stand the cold out at the edge of the floe, a small boat being required to pick them up when shot. These ducks will weigh five and six pounds apiece, their feathers making good beds.

Occasionally seal are shot but some sink before they can be picked up. Of these there are but three kinds known to the Eskimo. Kesegeer (or fresh water seal) – these seal are very dark with small light spots and are best looking for coats, though they are not as handsome as the same kind of seal in Hudson Strait. Others are known as nectyar and are smaller and lighter colored, darkest on the back, and their spots have more of a ring shape. The oujoug is the largest seal and its skin is of especial value for boot bottoms and making lines for various purposes. The women do not comb their hair for three days when one of these seal has been caught. There are a few walrus and one is occasionally caught, which is the principal dog food.

In the spring and fall once in a while a polar bear may be seen. Last fall (November) we got three which came near where we had some casks on shore, a large male, a female and cub. Had them skinned and salted to sell to some museum to have mounted when we get home. In March the she-bears come out of their winter quarters with their cubs. Wolves are not plentiful though we hear of them sometimes. A wolverine's tracks are sometimes seen near the vessel. White foxes are quite plentiful though in the summertime they have a blueish color. We have taken sixteen this winter and got twenty-eight others in trade in Hudson Strait. Reindeer were not plentiful this year as none of the natives have been very successful. They say that they must have kept away inland when they migrated south in the fall. Musk-ox are not found near here but to the north of us about 150 miles. We have five families off hunting them now. They went at Christmas time – expect them back about the first of April. We have taken (shot) quite a number of rabbits. They are large and pure white but the tips of their ears, which are black. They will weigh six to eight pounds apiece. Partridges are quite plentiful in the fall and spring, moving in large flocks and keeping out on the islands to keep clear of the foxes. How far north they go cannot say but do not lay around these parts. There are quite a number of small land birds in the spring. There are a few crows or ravens here during the summer and winter.

All the country about here is low and rocky and uneven with many fresh water ponds and these where they have a good depth of water are well supplied with salmon or salmon trout. Some of these ponds have outlets to the sea in summertime. From these the salmon find their way out in the spring to the salt water and back again in the fall. At such times are very thick in the streams. There is one such place at the head of this bay.

NATIVES: ESKIMOS (CAPE FULLERTON)

This part of the country between Wager River and Chesterfield Inlet does not seem to be the regular home of the Eskimo though they are now here and have come mostly because the vessels have made winter quarters between these two mentioned places. In the winter of 1893 and 1894 we wintered at Depot Island – barks *Canton, A.R. Tucker* – (which I do not consider a good harbor, being small and shallow, five fathoms, and quite difficult to get into). There were around us about 150 natives. Nearly if not all come from the vicinity of Repulse Bay and are known as Iwilic Inuies [Inuit]. Many of the men have two wives. Children are not plentiful as many of the women are barren, while one, and sometimes as many as four and five, but never have known or heard of more children in a family. I can say that these natives are very honest, though not very cleanly. After marriage (no ceremony) their faces, shoulder, wrists and lower limbs are tattooed. There is another tribe to the northwest known as Netchilic which are reported as given to thieving. The natives to the south are employed by the Hudson's Bay Company and are known as Kenepetu tribe.

USEFUL ARTICLES TO BRING UP

Iron rods one-quarter inch to one-half inch in diameter. Light bars of steel are also useful. These iron rods are used for seal spears. Large sized wire nails to be driven through the sled runners to prevent splitting (we afterwards in 1897–98 used stove bolts and think they are best as they are set up with a nut), and a large supply of smaller sizes as it takes many in building a house on the afterpart of the vessel to be used as a workshop and to move around in with some comfort.

FOR SAFETY OF THE VESSEL

It has always been a custom to cut the stem and rudder clear by the fifteenth of March and then when the vessel gets loose she can rise and not hurt either end. This winter we seem to have had a large amount of snow here (Cape Fullerton) while at Depot Island about forty miles to the west and south they have not had much more than half. I think it is caused by the northeast winds bringing the moist air from Roes Welcome. This large amount of snow has caused the ice to settle and at the same time to take the vessel with it, then as the water leaked up through, it made more ice on top while possibly the tides may have cut away the ice around the vessel underneath. The lifting power had become so great that we came up in our bed the nineteenth of February. We cut the rudder clear at once through over six feet of ice but so far as we could see by working the rudder it was alright. Would recommend keeping the rudder clear all winter.

Government expeditions 1903–5

(A) THE NEPTUNE EXPEDITION 1903–4

(i) Background and Command
(statement of A.P. Low)

The Dominion Government, in the spring of 1903, decided to send a cruiser to patrol the waters of Hudson Bay and those adjacent to the eastern Arctic islands; also to aid in the establishment, on the adjoining shores, of permanent stations for the collection of customs, the administration of justice and the enforcement of the law in other parts of the Dominion.

To perform these last duties, Major J.D. Moodie, of the Northwest Mounted Police, was appointed Acting Commissioner of the unorganized Northeastern Territories. Under his command were placed a non-commissioned officer and four constables of the Northwest Mounted Police, as a nucleus of the force that in the future would reside at these stations.

The *Neptune*, the largest and most powerful ship of the Newfoundland sealing fleet, was chartered as the most suitable vessel for the cruiser work ...

Early in June, 1903, I had the honour to be appointed, by the Honourable Mr. Préfontaine, the Minister of Marine and Fisheries, to the command of the expedition to Hudson Bay and northwards, on board the *Neptune*. I received instructions to proceed immediately to Halifax, to make necessary alterations to the ship, and to purchase all the provisions and outfit required for a two-years' voyage in the Arctics.[1]

1 A.P. Low, *The Cruise of the Neptune 1903–4*, 3, 4.

(ii) Instructions of Colonel F. White, Comptroller of the Royal North-West Mounted Police to Major J.D. Moodie

The Government of Canada having decided that the time has arrived when some system of supervision and control should be established over the coast and islands in the northern part of the Dominion, a vessel has been selected and is now being equipped for the purpose of patrolling, exploring, and establishing the authority of the Government of Canada in the waters and islands of Hudson Bay, and the north thereof.

In addition to the crew, the vessel will carry representatives of the Geological Survey, the Survey Branch of the Department of Interior, the Department of Marine and Fisheries, the Royal Northwest Mounted Police and other departments of the public service.

Any work which has to be done in the way of boarding vessels which may be met, establishing ports on the mainland of these islands and the introduction of the system of Government control such as prevails in the organized portions of Canada has been assigned to the Mounted Police, and you have been selected as the officer to take charge of this branch of the expedition.

You will have placed at your disposal a sergeant and four constables; you will be given the additional powers of a Commissioner under the Police Act of Canada, and you will also be authorized to act for the Department of Customs ...

It is not the wish of the Government that any harsh or hurried enforcement of the laws of Canada shall be made. Your first duty will be to impress upon the captains of whaling and trading vessels, and the natives, the fact that after considerable notice and warning the laws will be enforced as in other parts of Canada.[2]

(iii) Proclamation Distributed to Ships and Shore Stations, and Deposited in Cairns on Prominent Headlands through the Eastern Arctic

To Agents in Charge of Whaling and Trading Stations, Masters of Whalers etc., and all whom it may concern.

A detachment of the Northwest Mounted Police has been sent into Hudson Bay for the purpose of maintaining law and order and enforcing the laws of Canada in the territories adjacent to the said Bay and to the north thereof.

2 Quoted in A.E. Millward, ed, *Southern Baffin Island. An Account of Exploration, Investigation and Settlement During the Past Fifty Years* (Ottawa: King's Printer, Department of the Interior, North West Territories and Yukon Branch 1930), 14.

Headquarters have for the present been established at Fullerton. This has also been made a port of entry for vessels entering Hudson Bay and adjacent waters. All vessels will be required to report there and pay customs duties on dutiable goods before landing any portion of their cargoes on any place in the said territories.

Duty (on goods) imported into Canadian territories lying to the north of Hudson Bay will be collected for the present by a Canadian cruiser which will visit those waters annually or more frequently. Any violation of the laws of Canada will be dealt with by an officer of the police accompanying such cruiser.

By order.
J.D. Moodie
Commissioner of Police for Hudson Bay
and Territories to the North thereof.
(Fullerton is in N Lat 63°59',
w Long 89°20').[3]

(B) THE ARCTIC EXPEDITION 1904–5

Statement of Sir Wilfrid Laurier, Prime Minister, in the House of Commons

House in Committee of Supply.
For the purchase, equipment and maintenance of vessels to be employed in patrolling the waters in the northern portion of Canada; also for establishing and maintaining police and customs posts at such points on the mainland or islands as may be deemed necessary from time to time, $200,000.
SIR WILFRID LAURIER. Mr. Chairman, the committee, I am sure, has not forgotten that last year we sent an expedition to explore, patrol and assert the authority of the government of Canada in Hudson's [sic] Bay and the northern waters. The object of this expedition was fully explained to the House last year and met with general favour. The view was to assert beyond any possibility of doubt, so as to prevent any future possible conflict, the undoubted authority of the Dominion of Canada in the waters of Hudson's [sic] Bay and beyond. The steamer 'Neptune' was chartered last year and sailed from Halifax. It had on board a representative of the Geological Department, a representative of the Marine and Fisheries Department, Inspector Moodie and five men of the

---

3 Quoted in Diamond Jenness, *Eskimo Administration.* II *Canada*, Arctic Institute of North America, Technical Paper no 14 (1964), 19.

mounted police. The expedition for the time being was under the chief control of Mr. Lowe [sic]. He was to explore, as far as he could during the season, the northern waters and to establish a post somewhere in Hudson's [sic] Bay. No definite instructions were given to the expedition as to the location of the post. Then, as soon as the break of the ice in the spring would permit, the expedition was to go north and explore Baffin Bay and Lancaster Strait, and then come back to Cape Chidley, on the Straits of Hudson Bay. There they would be met by another boat, which was to sail, and has sailed on the 15th of July to meet them, give them coal, provisions, &c. The 'Neptune' is to come back and be relieved and be replaced by another boat, the 'Arctic', which will be under the command of Captain Bernier, and which is to sail on August 15. This boat will carry an officer and ten men of the mounted police, apart from the crew of the ship. They will relieve the 'Neptune'. Their instructions are to patrol the waters, to find suitable locations for posts, to establish those posts and to assert the jurisdiction of Canada. The government has been induced to come to this action because it is evident that the time has come when our interests in these northern waters should no longer be neglected. At the present time there are whalers and fishermen of different nations cruising in those waters, and unless we take active steps to assert, what is the undoubted fact, that these lands belong to Canada, we may perhaps find ourselves later on in the face of serious complications.[4]

4 Canada, *Official Report of the Debates of the House of Commons of the Dominion of Canada*. Fourth Session – Ninth Parliament, 4 Edward VII (Ottawa: King's Printer 1904), vol 67, cols 7968, 7969.

Glossary

This glossary consists of terms that may be unfamiliar to readers not addicted to the literature of whaling, the sea, and the polar regions. The names of Eskimo groups or 'tribes' are not included because they have been discussed in the Introduction.

Ambergris A fatty substance of great value occasionally found in the intestines of a sperm whale and used as a fixative in the perfume industry.

Anticoot An Eskimo spiritual ceremony directed by a shaman (see chap 2, n 6).

Backing (wind) A counter-clockwise change in the direction of wind origin (the opposite of hauling). For example, a northwest wind (one blowing from the northwest) could back to west.

Backstays Supports for a vessel's mast or masts, running down to fasten to the gunwales or deck astern of the mast(s).

Baleen (also called whalebone or bone) The filter-feeding material in the mouth of the Greenland whale and other non-toothed species. Approximately 700 slabs of baleen, weighing half a ton or more, hang from the upper jaw to prevent the escape of tiny krill which the whale takes in with sea water and swallows. Because of its lightness and resiliency baleen had many uses in industry and was much sought after by arctic whalers before 1915. The longest slabs sometimes exceeded twelve feet in length.

Banking A wall of snow blocks erected around a wintering vessel for protection against wind and cold. Also the act of erecting the wall.

Bark A square-rigged two- or three-masted sailing vessel with fore and aft sails on the mizenmast. The rig combined ease of handling with high manoeuvrability and was often used on whalers.

Barrel A wooden cask commonly used to transport oil in American whalers, with a capacity of 31.5 American or wine gallons.

Bay ice Landfast ice filling a bay, estuary, or fiord.

Bearded seal See ground seal.

Bend (a sail) To attach a sail to masts, forestay, yards or booms, ready for use. When wintering in the Arctic whalemen usually unbent the sails in the fall, stored them during the winter, and bent them again in the spring.

Berg The whaleman's usual term for iceberg.

Boat The whaleman's usual term for whaleboat

Boatheader The man in charge of a whaleboat – usually a mate – who steered the boat during the approach to the whale and had the responsibility of making the kill. As soon as the harpooner (or boatsteerer) had struck, the boatheader made his way forward to the bows to take up a lance or shoulder gun for the coup de grâce.

Boatsteerer (also called harpooner) The man who rose from his position at the forward oar to harpoon the whale from the bows, and who then moved aft to take up the steering oar or tiller while the kill was made by the boatheader.

Bolt (cloth) A unit of measure for canvas, approximately thirty-nine feet long and from two to three feet wide.

Bomb lance The explosive projectile fired from a shoulder gun (or bomb lance gun) to kill the whale after harpooning.

Bone The usual term for whalebone or baleen.

Boom A spar extending along the foot of a fore-and-aft sail, pivoted at the mast.

Braces Ropes running from the deck to the ends of yards, by which the crew could alter the angle of square sails to suit the wind direction and ship's course.

Brig A two-masted sailing vessel square-rigged on both masts.

Butt A large cask.

Cask A container constructed of wooden staves and headers, fastened with iron hoops, and used to carry ship's provisions, trade goods, oil, and a variety of other materials. Casks were made in various sizes (see barrel, butt, scuttlebutt, tierce).

Chain locker A space below deck up forward in which the anchor cables were stored.

Chronometer A ship's official timepiece, reliable even in rough weather, which permitted accurate determination of longitude.

Cooper The man responsible for assembling oil barrels from staves, and for carrying out general carpentry, including boat repair.

Cranes Timbers that swung out perpendicular to a ship's hull below the davits to support a whaleboat after it had been hoisted. To protect boats in rough weather the cranes were secured in a higher position and were then called upper cranes.

Darting gun A hand-thrown whaling weapon combining the functions of both harpoon and shoulder gun. The harpoon imbedded itself in the whale, attaching the animal to the whaleboat by a rope line. At the same time a small gun, triggered on impact, fired a projectile to explode inside the whale and kill it. This was especially useful in arctic regions, where harpooned whales (unless lanced and killed immediately) often dove under pack ice towing the whaleboats behind them.

Darting lance The explosive projectile for a darting gun, about sixteen inches long and one inch in diameter, with a time fuse activated on impact to detonate deep inside the whale.

Davits The curved supports of wood or metal which swing out from a ship's gunwales to lower or raise boats by means of ropes (falls) and blocks.

Deer The whalemen's name for caribou (*Rangifer arcticus*).

Duck A grade of canvas used in sails, bags, tarpaulins, and so on.

Falls The lower parts of the lifts which support a boom when the sail is lowered (cf runners). Also the tackles used to hoist and lower boats.

Fathom Six feet of linear measurement, used to describe water depth or the length of anchor cables.

Field ice See pack ice.

Flagging Presumably the cloth used for making flags, such as signal flags to direct whaleboats during a chase.

Flense To strip the blubber off a dead whale.

Floe edge (sometimes called the floe, or flaw) The seaward margin of the landfast or fast ice attached to the coast. Beyond the floe edge there is open water usually containing floating pack ice. In spring the floe edge retreats towards the shore as the fast ice melts and disintegrates.

Floes See ice floes.

Fluke The flat palm at the end of each arm of an anchor which digs into the bottom to hold a ship.

Fly A halyard on which a pennant, or narrow flag, is hoisted.

Flying jib One of a vessel's outer headsails (see appendix A).

Force (wind) The strength of the wind at sea according to categories on the Beaufort scale (see appendix D).

Foremast The mast closest to the bow, on which the foresail is set. On a schooner it is normally the shortest mast.

Foresail On a schooner the lower fore-and-aft sail on the foremast. A fore topsail could be set above it (see appendix A).

Fore topsail A sail set above the foresail, either between its gaff and the mast or hanging from a yard as on the *Era* (see appendix A).

Gaff A spar along the head of a fore-and-aft sail, its forward end pivoted at the mast.

Gallied Frightened (whales).

Gam A friendly visit and exchange of news between whaling ships at sea.

Ground seal The bearded seal (*Erignathus barbatus*) called oujoug (udjuk) by the Eskimos of the Canadian Arctic. A large animal weighing up to 800 pounds and resident in the Arctic all year. Its tough skin was utilized by the Eskimos for boot soles, dog traces, and umiaks.

Halyards Ropes used for hoisting sails extending from the head of a sail up the mast, around a block, and down to deck level.

Harpoon A wooden-shafted, hand-thrown weapon with a toggle iron point designed to turn sideways to prevent withdrawal from the whale's body. The harpoon was not intended to kill the whale but to attach it to the whaleboat by means of a line to maintain contact and tire it by the drag of the boat. The subsequent killing of the whale employed a hand lance or a shoulder gun.

Hauling (wind) A clockwise change in wind direction, the opposite of backing. For example, a northeast wind could haul to east.

House The cabin erected over the deck of a vessel when wintering in the Arctic, providing a useful space for social occasions, Eskimo trade, carpentry, and so on.

Iceberg Floating mass of ice calved into the sea from the snout of a glacier. Those found in the eastern Canadian Arctic usually originate along the coasts of Baffin Island, Devon Island, Ellesmere Island, or Greenland. Irregular in shape with about seven-eighths of their bulk below water, they are often transported long distances by ocean currents.

Ice cakes Small ice floes.

Ice floes Individual pieces of floating sea ice of various sizes.

Jib A triangular headsail mounted on a stay running from the bowsprit or stem to the foremast.

Jib-boom A spar extending beyond the end of the bowsprit.

Kedge A small anchor which can be carried by a small boat and used to shift the position of the vessel.

Ketch A small, two-masted, fore-and-aft rigged sailing vessel, with a large mainsail forward and a comparatively small mizensail aft.

Lance A long-handled weapon with which the boatheader killed harpooned whales (see also darting gun, shoulder gun).

Line tubs Open wooden tubs in the whaleboats containing the whale lines carefully coiled so that they could run out after the harpooned whale without snagging.

Loose-footed A sail without a boom along its foot.

Loose ice Pack ice open enough to sail through.

Log A contrivance which measures the distance of a ship's run each day; also a logbook.

Logbook The official written record of a ship's operation during a voyage, including information on course, distance, sail changes, winds, weather, navigational hazards, ports visited, whales killed, and so on. A whaler's logbook was normally kept by the first mate.

Mainmast The principal mast of a ship. In a schooner the mainmast is closest to the stern; in a ketch it is closest to the bow. The mast may be composed of several sections one above the other (such as the lower mast, topmast, and topgallant), depending on its height.

Mainsail The lower sail on the mainmast (see appendix A).

Main topsail On a schooner the upper sail on the mainmast, between the gaff of the mainsail and the mast (see appendix A).

Martingale A stay running from a vessel's stem to the end of her bowsprit.

Musk-ox (Ovibus muscatus) A heavy-coated animal not unlike the bison in appearance and size. After 1880, by which time few bison remained in North America, musk-ox skins were in great demand for winter coats and sleigh blankets.

Native boat Whaleboats owned and operated by the Eskimos (rather than the traditional Eskimo craft, the kayak and the umiak, which the whalemen usually referred to as 'skin boats'). The Aivilingmiut and Qaernermiut obtained used boats from the whaling ships through trade or employment. By 1903 there was an average of one whaleboat for every 2.5 families between Chesterfield Inlet and Repulse Bay.

Oujoug (or udjuk) Eskimo name for the ground seal.

Pack ice (also called field ice) An assortment of ice floes drifting beyond the margin of landfast ice, their concentration changing with winds and currents.

Partridge Ptarmigan (Lagopus lagopus; Lagopus mutus), one of the few species of bird that reside all year in the Arctic (see chap 3, n 21).

Pintals Vertical pins on the leading edge of a rudder which drop into circular gudgeons on the stern when the rudder is shipped.

Reef (sail) A reduction of sail area carried out by tying up a series of cords (reef points) along the lower part of the sail.

Run The bottom part of a ship's stern section and the cargo space inside it.

Runners The upper parts of the lifts which support a boom when the sail is lowered (cf falls).

Saddle (caribou) The section of the caribou carcass between front and hind quarters.

Saleratus Baking soda.

Scantling Small pieces of lumber, probably used as studs and rafters in constructing the winter house over the deck of the Era.

Schooner A fore-and-aft rigged sailing vessel containing two or more masts with the rear mast (mainmast) carrying the largest sail (see appendix A).

Scurvy A potentially fatal disease resulting from a deficiency of vitamin C. It was often experienced by seamen on long voyages during which there was insufficient consumption of fresh meat, vegetables, and other anti-scorbutics.

Scuttlebutt Water cask.

Sharpie A small flat-bottomed boat. In Hudson Bay sometimes built on board whalers and used to retrieve seals shot by rifle from the floe edge.

Sheets Ropes running from the clew of a sail (its rear, lower corner) or its boom to the deck and used to control the set of the sail.

Ship In its specific meaning, a sailing vessel with three or more masts, square-rigged on all masts.

Shoeing The sliding layer beneath the wooden runners of a dog sled, usually made of whale jaw-bone or more recently of iron.

Shooting lance Presumably the projectile or bomb lance fired from a shoulder gun.

Shoulder gun A hand-held gun weighing from nineteen to twenty-seven pounds which fired a small explosive projectile (bomb lance) into a whale, normally a whale already harpooned, exhausted by the chase, and weakened by loss of blood. The Brand gun, invented in 1850, was a muzzle-loader. The more sophisticated Eggers gun was breech-loading (cf lance, darting gun).

Sidelights The red and green navigational lights carried at night by ships on their port (left) and starboard (right) sides respectively.

Slop chest The store of clothes and other useful articles issued periodically to the crew on credit (see appendix C).

Slush To grease a vessel's wooden masts after scraping.

Smack A small coastal fishing vessel, usually single-masted.

Soapstone A soft talcose rock, traditionally used by the Eskimos for making lamps and pots.

Spades Long-handled knives used to cut whale blubber.

Spermaceti A waxy solid which separates from the oil in the head, or case, of a sperm whale and highly valued for use in candles and lubricants.

Square sail A four-cornered sail hanging from a yard across the vessel. (The opposite of a fore-and-aft sail, which is usually attached at its leading edge to a mast or stay and more in line with the ship's hull than at right angles to it.) The rig of a schooner is fore-and-aft, but the *Era* could set two square sails on foremast yards, to make the most of following winds (see appendix A).

Steerage The section aft in a whaler occupied by mates, cooper, carpenter, and steward. Ordinary seamen lived in the forecastle.

Strake A plank running end to end in a wooden hull.

Sumner method A navigational procedure invented in 1837 to determine a ship's position from a position line.

Tack To sail a vessel into the wind in a zig-zag pattern, the ship being first on the starboard tack (wind from the starboard, or right, side) and then on the port tack. In coming about from one tack to another the ship swings her bow across the direction from which the wind is blowing (the opposite of wear).

Tierce A cask, usually with a capacity of forty-two wine gallons, often used to store ships' provisions, such as salt meat, bread, and molasses.

Toggle iron The sharp blade of a whaling harpoon, shaped and pivoted to turn at right angles to its metal shaft after penetrating the skin.

Topgallant Normally the highest section of a ship's mast, and the square sail mounted upon it (see appendix A).

Topsail A sail mounted above a gaff or on square yards attached to the topmast (see appendix A).

Throat The part of a fore-and-aft sail where its gaff joins the mast.

Tryworks Brick-encased melting pots on the upper deck for rendering blubber into whale oil (trying out). The blubber scraps from the pots were fed as fuel into the fire beneath. After cooling, the oil obtained was run below into barrels.

Tubs See line tubs.

Umiak An open Eskimo boat made of sea mammal skin stretched over a wooden frame, normally between twenty-five and thirty-five feet in length.

Wear To change a ship's course from one tack or reach to another by turning the stern across the wind (the opposite of tack).

Whaleboats Small, open, double-ended boats from which whales were pursued, harpooned, and lanced. They were approximately thirty feet long, propelled by five oars or by sail, and steered either by rudder and tiller or by a long steering oar, or sweep, over the stern. Because of their seaworthiness, versatility, durability, and capacity, they were greatly prized by the Eskimos of northwestern Hudson Bay (see native boats).

Whalebone See baleen.

Whale bones The bones of whales (not the same as whalebone or baleen).

White whale (or beluga; *Delphinapterus leucas*). A small whale approaching twenty feet in length.

Wolverine (*Gulo luscus*) A strong and crafty four-footed animal about three feet long whose fur was prized for the trim around parka hoods.

Yard A spar crossing a ship's mast (usually horizontal and at right angles to the fore-and-aft line of the vessel's hull) from which a square sail can be set. The *Era* carried yards on her foremast (see appendix A).

Selected Bibliography

Bernier, J.-E. *Master Mariner and Arctic Explorer: A Narrative of Sixty Years at Sea from the Logs and Yarns of Captain J.E. Bernier*. Ottawa: Le Droit 1939

Boas, Franz 'The Eskimo of Baffin Land and Hudson Bay. From Notes Collected by Capt. George Comer, Capt. James S. Mutch, and Rev. E.J. Peck.' *Bulletin of the American Museum of Natural History*, vol 15, part 1 (1901): 1–370. Facsimile reprint. New York: AMS Press 1975

– 'Second Report on the Eskimo of Baffin Land and Hudson Bay. From Notes Collected by Captain George Comer, Captain James S. Mutch, and Rev. E.J. Peck.' *Bulletin of the American Museum of Natural History*, vol 15, part 2 (1907): 371–570. Facsimile reprint. New York: AMS Press 1975

Borden, K. Ethel 'Northward 1903–4.' *Canadian Geographical Journal*, vol 62, no 1 (1961): 32–9

Borden, Lorris Elijah 'The Lost Expedition, Being the Diary of L.E. Borden, Surgeon and Botanist with the First Canadian Government Expedition to the Hudson Bay and Arctic Islands, as Recorded on Board D.G.S. *Neptune*, 1903–4.' Typescript. Ottawa: Public Archives of Canada (MG 30 C.52, vol 2)

– 'Memoirs of a Pioneer Doctor.' Part 3. Typescript. Ottawa: Public Archives of Canada (MG 30 C.52, vol 2)

Comer, George 'Whaling in Hudson Bay, with Notes on Southampton Island.' In B. Laufer, ed, *Boas Anniversary Volume: Anthropological Papers Written in Honor of Franz Boas*. New York: Stechert 1906, 476–84

– 'A Geographical Description of Southampton Island and Notes Upon the Eskimo.' *Bulletin of the American Geographical Society of New York*, vol 42 (1910): 84–90

– 'Additions to Captain Comer's Map of Southampton Island.' *Bulletin of the American Geographical Society of New York*, vol 45 (1913): 516–18

- 'Notes by G. Comer on the Natives of the Northwestern Shores of Hudson Bay.' *American Anthropologist*, n.s., vol 23 (1921): 243–55
Dorion-Robitaille, Yolande *Captain J.E. Bernier's Contribution to Canadian Sovereignty in the Arctic*. Ottawa: Indian and Northern Affairs 1978
Eifrig, C.W.G. 'Ornithological Results of the Canadian *Neptune* Expedition to Hudson Bay and Northward 1903–1904.' *Auk*, vol 22, no 3 (1905): 233–41
Fairly, T.C. *The True North: The Story of Captain Joseph Bernier*. Toronto: Macmillan 1954
Ferguson, Robert *Arctic Harpooner: A Voyage on the Schooner Abbie Bradford, 1878–79*. Edited by Leslie Dalrymple Stair. Philadelphia: University of Pennsylvania Press 1938
Halkett, Andrew 'A Naturalist in the Frozen North.' *Ottawa Naturalist*, vol 19, no 4 (1905): 79–117
Low, A.P. *The Cruise of the Neptune, 1903–4: Report on the Dominion Government Expedition to Hudson Bay and the Arctic Islands on Board the D.G.S. Neptune, 1903–1904*. Ottawa: Government Printing Bureau 1906
McGrath, P.T. 'A new Anglo-American Dispute: Is Hudson Bay a Closed Sea?' *North American Review*, vol 177 (1903): 883–96
- 'Whaling in Hudson Bay.' *New England Magazine*, n.s., vol 30, no 1 (1904): 188–98
Minotto, C. 'La Frontière arctique du Canada: Les Expéditions de Joseph-Elzéar Bernier (1895–1925).' Master's thesis, McGill University, Montreal 1975
Moodie, J.D. 'Copy of Daily Diary' [1903–4]. Ottawa: Public Archives of Canada (RG 18 A.1, vol 281, file 716)
- 'Report of Superintendent J.D. Moodie on Service in Hudson Bay, per SS *Neptune*, 1903–4.' Ottawa: House of Commons (4–5 Edward VII, Sessional Paper no 28) 1905, 3–12
- 'Report of Superintendent J.D. Moodie on Service in Hudson Bay (per SS *Arctic*, 1904–5).' Ottawa: House of Commons (5–6 Edward VII, Sessional Paper no 28) 1906, 2–16
RCMP Records Documents Relating to the Administration of the Royal Canadian Mounted Police: Comptroller's Office, Official Correspondence 1874–1919. Ottawa: Public Archives of Canada (RG 18 A.1)
Ross, W. Gillies *Whaling and Eskimos: Hudson Bay 1860–1915*. Ottawa: National Museum of Man (Publications in Ethnology, no 10) 1975
- 'Canadian Sovereignty in the Arctic: The *Neptune* Expedition of 1903–04.' *Arctic*, vol 29, no 2 (1976): 87–104
- 'Whaling and the Decline of Native Populations.' *Arctic Anthropology*, vol 14, no 2 (1977): 1–8

- 'The Annual Catch of Greenland (Bowhead) Whales in Waters North of Canada 1719–1915: A Preliminary Compilation.' *Arctic*, vol 32, no 2 (1979): 92–121

Stackpole, Renny A. *American Whaling in Hudson Bay, 1861–1919*. Mystic, Conn: Munson Institute of American Maritime History, Marine Historical Association 1969

Vanasse, Fabien 1905. 'Relation sommaire du voyage de l'*Arctic* à la Baie d'Hudson, 1904–5.' Typescript. Trois Rivières: Archives du Seminaire

Zaslow, Morris *The Opening of the Canadian North 1870–1914*. Toronto: McClelland and Stewart (Canadian Centenary Series) 1971

Index

A.R. Tucker 27, 54 n16
A.T. Gifford 219, 223
Abbie Bradford 28, 32
Active 51 n11, 54 n17, 55, 56 n20, 74 n13, 125 n8, 130, 133, 139, 140, 207
Aivilingmiut (Iwillic) 20, 21, 32, 40 n2, 51, 79, 95
Alaska boundary dispute (1903) 35
Alcohol 22, 42, 45
American Museum of Natural History 28, 31, 114 n, 138, 219
Amundsen, Roald 176, 178
Anticoot 69, 82, 85, 88, 171, 174, 182
Arctic 146, 149 n, 152 n, 154, 157, 158, 159, 163
Arctic Bay 220
Arctic Islands 34

Baffin Island 220
Baker Lake 63 n2, 157, 158, 163, 179, 221
Baleen see Whalebone
Ballast 188, 209
Banking (snow) 76, 77, 80, 117, 154, 155, 156, 157, 160, 161, 162, 181
Bartlett, M. 94

Bartlett, Sam W. 62, 65, 69, 86, 91, 105, 108, 115
Beach Point 54, 57, 176, 200, 209
Beacon Island 60 n, 195
Bears: skins 51, 58, 64, 82, 115, 117, 118, 140, 177, 185, 188, 206, 213; killed, 99, 114, 126, 128, 129, 180, 196, 198, 204, 207; sighted, 99, 129, 158
Bering Strait 177
Bernier, Captain Joseph-Elzéar 148, 155, 157, 158, 159, 165, 166, 170, 171, 173, 174, 182
Birds: partridge 113; snow bunting 113; hawk 118; eider ducks 126, 128, 189, 190, 191, 196; snow 180; seagulls 118, 190, 205; 'south-southwest' ducks 190
Blue Lands 57, 138, 139, 201, 204
Boas, Franz 41 n4, 74 n14
Borden, Dr Lorris Elijah: injuries 71, 72; and *Era* 72, 89, 90, 91, 92, 94, 98, 102, 104, 105, 106, 107, 108, 110, 117; and Shoofly 151–2, 152–3; and Moodie 75 n16; and Comer 83, 96; lectures 90; and

carvings 111; and Southampton Island 122, 124 n, 127 n
Bushman Island 202, 208

Caldwell, G.F. 94, 96, 109, 146 n
Canada, government of 149, 220
Canadian Polar Expedition 148, 149 n
Canton 26–7
Cape Edwards 202, 204, 205, 208
Cape Fullerton 8, 51, 58, 59, 60, 67, 117, 122, 133, 139, 140, 141, 210, 211
Cape Jalabert 120 n5, 121
Cape Kendall 51, 128 n
Cape Low 128 n
Cape Martineau 202 n, 205, 206
Caribou Eskimos 15, 16; see also Eskimos
Carvings: by Comer 85, 155, 157, 160, 163, 165, 169; by Eskimos 88, 90, 92, 180, 181
Chapel, Christopher 7
Chapel, Edward 7
Chesterfield Inlet 62, 65–6, 91, 109, 120, 176
Chronometer 84, 87, 134, 216
Cleveland, George Washington 57, 133, 140
Clothing 71, 78, 97, 115, 116, 132, 137, 143, 145, 185
Coats Island 50, 51, 112 n20, 211
Comer, Captain George: journal 13, 29, 30–3, 39; and Eskimos 22, 30, 55, 56, 57, 74, 127, 171, 174, 191; youth 22–3; arctic voyages (1889–1902) 23–4, 25–8; antarctic voyages (1879–89) 24–5; scientific collecting 28, 127, 128, 138, 196, 197, 198, 207; and Neptune 38, 61–2, 63, 67, 104; and graphophone 40, 51, 52, 60, 84, 92, 93, 97, 99, 100, 137, 142, 143, 145, 156, 187; and casts of Eskimo faces 41, 88, 89, 90, 91, 99, 167, 178, 179, 184; photographs 40 n2, 41, 47, 52, 71, 73, 84, 85, 91, 92, 95, 120, 167, 168, 173, 174, 175, 176, 178, 181, 182, 186, 187, 188, 193; health 43, 48, 49, 69, 72, 73, 105, 106, 117, 121, 122, 123, 171, 172, 205; and sick 55, 60, 75, 77, 100, 170; and Neptune 61–2, 67; and newspaper 69, 77, 156, 160, 169; and Eskimo songs 73, 88, 93, 120, 167, 177, 221 n; and regulations 74, 75, 77, 78; and Low 80, 83, 111; and Moodie 75 n16, 98, 106 n14, 119 n, 178–9; and crew 103, 104, 164, 170–1; and Borden 111, 117, 118; and Southampton Island 123–31, 192–9; and Lyon Inlet 133–9, 201–9; carvings 160, 163, 165, 169; and Bernier 166, 170, 174; and whales 194, 203; life 218–20; later voyages 218, 219; midden 219
Comer Strait 134 n
Crew: experience 42, 43, 45, 93; health 42, 43, 45, 46, 47, 59, 60, 75, 88, 89, 91, 93, 94, 98, 102, 104, 105, 107, 110, 111, 115, 117, 120, 140, 141, 164, 165, 170, 171, 175, 206, 215; behaviour 53, 76–7, 103, 104, 164, 170–1; complaints 86, 103, 106; sports 101, 106, 166, 170; duties 158
Customs duties 61, 73, 87, 94

Dances: on Neptune 75, 78, 80, 82, 84, 86, 88, 89, 90, 91, 93, 107; on

Era 78, 79, 82, 87, 88, 93, 96, 97,
102, 107, 110, 111, 116, 150, 154,
158, 160, 163, 164, 171, 172, 173,
174, 181, 183, 185; ashore 142; on
Arctic 153, 154, 156, 159, 161,
162, 165, 168, 169, 170, 171, 173,
176, 180, 181, 182, 184, 186
Deaths 80–1, 93, 107, 112, 192, 198
Depot Island 8, 60, 65, 67, 83, 85,
105, 114, 121, 180
Disease 51, 56 n, 91
Duke of York Bay 134, 207, 208

Ellis, Richard L. 43, 59, 60, 150, 175,
203, 205, 206
Electricity 154
Entertainment 70, 167, 170, 185; see
also Dances
Era: trading outfit 19; arctic
voyages 25–8; condition 43, 45,
47, 69, 79, 80, 81, 83, 85, 87, 89, 90,
91, 93, 94, 97, 98, 101, 102, 106,
110, 111, 112, 116, 117, 118, 119,
133, 144, 148, 151, 155, 161, 165,
169, 172, 180, 189, 199, 204, 210,
212, 215, 217; maintenance 43, 59,
61, 63, 93, 131, 144, 157, 171, 174,
185, 187, 189, 199, 201, 212, 213,
214, 215, 216; wrecked 211 n, 218
Ernest William 14, 54, 207, 208
Eskimos: and whaling 6, 14, 16-18,
21, 204, 220–1; employment 8,
14, 18–19, 22, 52 n13; groups 14–
17; and European goods 17–21;
and guns 19–20, 220; and
whalemen 21–2, 151 n; burial
practices 57, 192; and *Era* 58,
91, 111, 118, 120, 174, 183; and
snow houses 82, 83, 115, 152, 153,
167, 183; taboos 93, 156, 167,

192; and fishing 96, 97, 99, 170,
186; on Southampton Island 134;
Thule Eskimo culture 135 n, 136,
219 n; and medicine 163, 174;
starvation 184; Dorset Eskimo
culture 219 n; population 220–1;
recent trends 221; see also
Aivilingmiut; Caribou Eskimos;
Comer; Netsilingmiut; Padlimiut,
Qaernermiut; Sadlermiut;
Sauniktumiut; Sinimiut;
Tununermiut

Farhill Point (Igloo-ju-ack-talic) 135,
202 n
Faribault, Dr G.B. 72, 102, 112
Ferguson River 98, 99
Ferguson, Robert 28, 32
Fisher, Captain Elnathan B. 26–7
Fisher Strait 26, 112 n20, 128
Flood, Dr 160, 162, 163, 165, 171,
173, 175, 187, 191
Forman, William 201
Fort Churchill 18, 81, 167, 168, 173,
182
Fox, arctic 153, 163, 185
Foxe Basin 55 n19, 221
Francis Allyn 24, 57 n22, 106
Frobisher, Martin 221
Frozen Strait 134, 137, 138, 201, 207,
208, 222
Fullerton Harbour 27, 60 n23, 87,
110 n, 112, 116 n, 118 n, 130, 141,
143, 199
Fur trade 13-14, 63 n2
Furs and skins 19, 38, 51, 52, 53, 57,
58, 59, 64, 71, 73-4, 75, 77, 78, 79,
82, 98, 100, 109, 114, 117, 118, 130,
140, 144, 153, 154, 157, 158, 159,
162, 169, 170, 173, 177, 185, 191, 219

Game   60, 61, 63, 64, 70, 71, 72, 73, 77, 78, 79, 82, 83, 84, 85, 87, 89, 90, 91, 92, 93, 94, 95, 98, 99, 101, 102, 104, 107, 109, 111, 112, 114, 121, 122, 130, 141, 142, 144, 148, 151, 152, 153, 154, 155, 156, 158, 159, 160, 162, 164, 166, 167, 170, 171, 172, 174, 175, 177, 178, 179, 180, 181, 183, 189, 192, 197, 198, 200, 208

George B. Cluett   219

Gjoa   176, 178

Gore Bay   55, 56, 135, 137, 202, 204, 206, 207

Graphophone   40, 51, 52, 60, 73, 84, 88, 92, 93, 97, 98, 99, 120, 137, 142, 145, 156, 167, 177, 187; see also Comer

Ground seal (oujoug)   93, 94, 107, 116, 117, 121, 126, 127, 128, 129, 132, 196

Halkett, A.   87–8

Hall, Charles Francis   221

Hanbury, David   176 n4

Harbours, winter   8, 16

Harpoon   9

Harvaqtormiut   16

Harwood, P. LeRoy   31

Hauneqtormiut   16

Hayne, Staff-Sergeant (RNWMP)   175, 191, 200

Hearn, J.   64, 72, 84, 94, 101

Herschel Island   36

Houses: at Whale Point   58; at Fullerton Harbour   91; at Repulse Bay   139; over deck   63, 64, 68, 90, 104, 113, 143, 144, 152, 188, 189; ice-slab   153; sold   189

Hudson Bay   36–8; see also Whaling

Hudson's Bay Company   7, 37, 38, 168 n

Hurd Channel   134–5, 136, 137, 207, 208, 222

Ice: pack   49, 50, 120, 190; pond 67 n4, 70, 73, 76, 149, 151, 158, 185, 186, 187; for Eskimo houses   152, 153

Icebergs   44, 46, 47, 48, 212, 213, 214

Iglulik Eskimos   15, 16

Iglulingmiut   41 n4, 51, 221

Inuit   see Eskimos

Iwilic   see Aivilingmiut

Kenepetu (Qaernermiut)   16

Knight, James   17

Labrador   96

Laurier, Sir Wilfrid   37, 152 n, 182

Leden, Christian   219

Lopes, Brass   48, 65, 90, 104, 105, 158, 194

Low, Albert Peter: and Comer   65, 76, 80, 81, 83, 84, 89, 106, 109, 110, 141, 142; and *Era*   83, 92, 102, 108, 109, 110, 118–19; lectures   89, 96; and carvings   90, 111; explorations   96 n5, 112, 114, 117, 122, 124; scientific collecting   124 n, 127 n; and Eskimos   141, 142; and Moodie   146 n

Luce, Charles T.   37

Luce, Thomas   40, 41, 217

Luce, Thomas and Sons   38, 54 n16, 57 n22, 109 n, 140, 211 n

Lyon Inlet   54 n17, 135 n

Mackean (artist on *Arctic*)   160, 164, 165, 180, 184

Mail 55, 58, 81, 130, 139, 167, 168, 182, 208, 210, 211, 216
Manico Point 125, 127, 196
Mansel Island 50, 211
Marble Island 8
Maynes, W.P. 107, 109
Medical treatment 48, 60, 72, 75, 77, 93, 94, 96, 106, 153, 183, 100 n8, 106 n14
Moodie, A.D. 147 n2, 154, 157, 158, 163, 168
Moodie, Grace 147, 163, 170, 186
Moodie, Superintendent J.D.: and trade 63 n2, 73-4, 75, 87, 94, 98, 163; house 67, 69, 102, 157; powers 72 n12, 103 n12, 146 n, 149 n, 177 n; and Eskimos 75; character 75 n16, 106, 150, 172, 177; lectures 91; and Comer 103, 106, 108, 114, 118 n, 164; and *Era* 131, 147, 150, 165, 168, 186, 189; and *Neptune* 146 n; lantern slides 169, 187; and Amundsen 177–8, 179
Mosquitoes 131, 198, 199
Murray, Captain Alexander 126 n, 139, 140, 201 n
Murray, Captain John W. 109, 133, 201 n
Musk ox: skins 51, 52, 53, 57, 58, 59, 71, 79, 109, 117, 118, 130, 140, 144, 191; trade 73-4, 75, 78, 98; meat 114; heads 114, 115; Royal Commission 219

Natives: Ben (Arb-lick) 22, 179, 181, 189, 190, 192; Blockhead (Artung-e-lar) 176, 178, 179; George (Kan-ne-uke) 114, 184, 188; Gilbert 66, 179, 202; Harry (Teseuke) 22, 54, 90, 91, 129, 179; Jack 82, 190, 206; Jimmy 114, 140, 169, 181; John L. 167, 168, 169; Keckley 79, 190; Melichi 22, 179; Mike 165, 168; Tom Nolyer 68, 110, 170; Paul 56, 91; Sam 66, 88, 179, 195; Shoofly 151, 152–3, 155, 156; Smiley 30, 210, 211; Stonewall 69, 88, 90, 102, 105, 114, 184; Tom Luce 94
Natives, ship's 51, 52 n13, 53, 79 n18, 91, 117, 140, 174, 177, 209
Navigation 43, 44 n6; 45, 46, 47 n, 84, 87, 112 n20, 134, 155, 213, 214, 217
Neptune 37, 61, 65–7, 79 n18, 130, 141, 142, 146 n
Neptune expedition 37–8, 63, 64, 92 n, 104
Netsilingmiut 15, 16, 17, 21, 28, 32, 41 n4, 51, 201
New Bedford (Mass) 40, 54 n16, 217, 218
Newfoundland 38, 211 n
New London (Conn) 40, 186
Newspapers 70, 77, 155, 169
Nile 23–4, 114, 186
Northwest passage 176 n3

O'Connell, Frank 80, 81

Padlimiut 18, 41 n4
Parry, William Edward 18, 27, 202 n
Peary, Robert 35
Peck, Rev E.J. 141 n20
Photography 95, 108, 110, 186; see also Comer
Provisions 46, 52, 59, 69, 73, 75, 79, 86, 88, 89, 91, 96, 97, 99, 102, 114, 119, 122, 130, 131, 137, 145, 150,

156, 169, 170, 173, 181, 184, 186, 187, 188, 190, 192, 196, 200, 204, 208, 212, 217

Qaernermiut (Kenepetu): and whaling   16, 32; nomenclature   17; and guns   20; and whaleboats   20; at whaling harbours   21; at ships   64, 71, 79, 100, 143, 144, 177; photographs   40 n2, 95, 188; plaster casts   41 n4; deaths   51; songs   73; and *Neptune*   79 n18
Quarmats   82 n21

Rae, John   18
Repulse Bay   51 n11, 54 n17, 126, 133 n, 135 n
Resolution Island   46, 47, 48
Reynolds, Herbert R.   158, 193, 194, 202, 205
Roes Welcome Sound   53, 116 n, 139, 140, 200, 222
Ross M.   86, 105
Royal North-West Mounted Police   61 n25, 67, 69, 72 n12, 102, 120 n4, 147, 150, 152 n, 189 n, 198, 199, 211 n
Rupert's Land   34, 38

Sadlermiut   16, 21, 28, 51, 95, 125, 127, 196, 197, 198
Sauniktumiut   41 n4
Scurvy   92 n, 104, 105, 107 n, 206
Sea elephants   24
Shamans (Angakoks)   69 n6; see also Anticoot
Shark, Greenland   205
Show-vock-tow-miut   17, 167
Sinimiut   15, 16, 41 n4
Sleds   72, 157

Snow: for banking   76 n17, 156, 157, 159, 162; for Eskimo houses   82, 83, 152, 167, 183
Soapstone   15, 101
Southampton Island   56 n20, 122 n, 124 n7, 126, 128 n, 134
Spicer, Captain J.O.   23, 25–6, 114, 186
Spicer Island station   14, 25, 49 n10
Stefansson, Vilhjalmur   29–30
Sverdrup, Otto   35

Tents   21, 143
Tobacco   48, 78, 97, 167, 203
Trade: ivory   13, 19; values   33 n, 179
Tryworks   10, 185, 187
Tucker, Charles   164, 165
Tununermiut   15, 95

Umiaks   18

Vanasse, Fabien   177 n, 184
Vansittart Island   137, 138, 202 n, 206, 207, 208

Wager River   90 n3
Walrus   92, 94, 97, 99, 120, 172, 179, 192, 200
Water, drinking   52, 53, 67 n4, 132, 139, 209
Weather observations   98 n6, 99 n7
Whaleboats: characteristics   9; cruises   11–12, 116 n, 118, 119, 120, 130, 132, 186, 189; importance   19–20; work on   47, 107, 108, 111, 112, 113, 114, 115, 117, 122, 129, 132, 174, 181, 182, 183, 184, 193, 195, 196, 199; covers   52, 53, 206; shoeing   52, 59, 109, 110;

numbers   53, 55, 117, 132, 139, 205;
  at floe edge   116, 189; capsized   194
Whalebone: uses   4, price   12;
  theft   54, 78; splitting and
  scraping   78, 132, 195, 203, 205;
  cutting out   194, 195, 202, 203, 205,
  206; weighing   200, 204
Whale Point   51, 53 n15, 124, 163
Whales: Balaena mysticetus   4;
  flensing   6, 10; killed by *Era*, 51,
  54, 129, 194, 195, 198, 202, 203, 205;
  killed by Scots   54, 57, 137, 203,
  207; white   102; finback   114;
  sighted   125, 128, 138, 139, 193,
  194, 196, 201, 202, 203, 205, 206,
  207; lost   138, 202, 203; deple-
  tion   138 n17, 222, 223;
  carcasses   204; feeding   206 n12;
  reproduction   222
Whaling: Arctic   3, 4, 13, 14, 29, 32;
  Davis Strait   3–4, 5–7, 13; Hudson
  Bay   4, 7–12, 13, 36; West Side
  5–7; Cumberland Sound   5–7;
  methods   8–12; weapons   9–11;
  floe   6–7, 11, 52 n12, 116 n;
  spring   11–12, 107, 108, 109, 110,
  111, 112, 113, 114, 115, 116 n, 174,

181, 182, 183, 184, 185, 186;
  logbooks   29; gear   46, 48, 52, 76
  n16; prospects   137; and
  Eskimos   220
Whaling stations   14, 74 n13;
  Blackhead Island   25–6, 74 n13;
  Cape Haven   25–6, 73 n13;
  Kekerten Island   74 n13;
  Southampton Island   14, 51 n, 54
  n17, 57, 125, 126, 129, 196; Spicer
  Island   14, 25, 49 n10; Wager
  Bay   14, 57, 139, 140
White, Fred   72 n12, 152 n, 211 n
Wintering   6, 8, 10–11, 52, 61, 63, 64,
  68, 69, 76, 101, 104, 112, 113, 117,
  118, 130, 141, 143, 144, 145, 148,
  149, 151, 152, 154, 155, 156, 157,
  162, 181, 184, 189
Winter Island   202 n, 205
Wolverines   53, 64, 79, 114, 118, 140,
  158, 169, 173, 185
Wolves   52, 53, 57, 64, 71, 100, 106,
  114, 118, 140, 177, 185

Yellow Bluff   58, 82, 140, 180, 193,
  210